Hum Factors
for General Aviation

D0342348

Stanley R. Trollip, Ph.D.
Richard S. Jensen, Ph.D.

Published by
JEPPESEN SANDERSON

No part of this publication may be reproduced, stored in a retrieval system, or transmitted in any form or by any means, electronic, mechanical, photocopying, recording, or otherwise without the prior permission of Jeppesen Sanderson, Inc.

JS319005A

To Dr. Stanley N. Roscoe, mentor and friend.

Table of Contents

Acknowledgements

We want to acknowledge our debt and express our gratitude to all those who have made this book possible.

For his vision, guidance, and mentoring, as well as his friendship, both of us thank Dr. Stanley Roscoe, former Director of the Aviation Research Laboratory at the University of Illinois. He continues to be a major influence in communicating the study of human factors for the benefit of general aviation. Both of us have also benefitted from our colleagues in the field, too numerous to mention, who have written, argued, and debated the topics we cover in the book. Their thinking has influenced us greatly.

We also want to acknowledge Frank Hawkins and his book *Human Factors in Flight*, and David O'Hare and Stanley Roscoe and their book *Flightdeck Performance: The Human Factor*. We used both books as resources and adapted ideas and illustrations for this book.

We also want to thank Randy Fogerty, Richard Reinhardt, Eve Sheridan, and Dan Wallace for their valuable advice on improving the manuscript. Our thanks also to Lynn Brantley for her assistance in locating scenarios and illustrations for the book. Finally, we have been extremely fortunate to have had active and informed feedback from the editors at Jeppesen Sanderson. Their input has also improved the manuscript.

Preface

In 1986, the Assembly of the International Civil Aviation Organization (ICAO), adopted Resolution A26-9 on Flight Safety and Human Factors. As a follow-up to this Resolution, the Air Navigation Commission formulated the following objective for the task:

> To improve safety in aviation by making States more aware and responsive to the importance of human factors in civil aviation operations through the provision of practical human factors material and measures developed on the basis of experience in the States.

We wrote this book partly in response to the ICAO call for materials for use in civil aviation and partly because both of us are acutely aware of how easy it is to make errors, even when aware of the traps a pilot can so easily fall into.

Over 80% of the accidents in general aviation are caused by pilot error. It is the actions, reactions, and decisions of pilots that cause most accidents — not catastrophic failures of aircraft systems, or air traffic control errors.

All over the world, students learn to fly by studying the airplane and its systems, the weather and its effect on flying, and the rules and regulations governing flight. Seldom do they study the central and most critical component of any flight, namely human beings in the cockpit. Seldom are they provided the tools to help them avoid making the common mistakes that cause accidents.

Recently, the International Civil Aviation Organization has recommended that training organizations provide such tools, largely through an increased emphasis on human factors. This book responds to that recommendation. It is about human beings in the cockpit and the human factors that affect their ability to fly safely. We discuss major issues that influence how well pilots fly; issues that are both inherent in pilots' mental and physical states, as well as those caused by the environment. We devote chapters to each of the following topics:

WHY ACCIDENTS HAPPEN

In the introductory chapter, we provide information about accident statistics and some of the reasons why accidents happen. We introduce a model that describes pilot performance, which we use as a basis for the chapters that follow.

LEARNING TO FLY

All pilots go through a learning process, both for the academic side of flying (ground school) and for the skills of controlling an airplane (flight training). We discuss a variety of techniques that will help you learn more effectively, and we provide some guidance to flight instructors on the fundamental factors for good instruction. We also provide guidelines on how to evaluate flying schools.

COCKPIT DESIGN

Accidents are often attributed to pilot error. This chapter shows that it is often not the fault of the pilot, but more so of the design of instruments and controls. We give examples of both good and poor design.

THE EYES AND EARS

A pilot's eyes and ears provide the primary conduit for information needed to fly an airplane. Often, what pilots think they see is not what is actually there. Their perception can be influenced by visual illusions, preset notions of what should be seen, and other psychological and physiological factors. This chapter covers these topics, so that you may be more aware of how your perception is influenced.

THE BRAIN

All information you receive is processed by the brain as you fly the airplane, make decisions, and react to changes in the environment. In this chapter, we briefly introduce the parts of the brain, their roles, and how they influence our thinking.

THE BODY

The physical well-being of your body influences how well you are able to perform. We discuss how your body reacts to oxygen deprivation, substance abuse, diet, and stress, with particular reference to your performance as a pilot. We also suggest some techniques that can improve your overall physical state.

EMOTIONAL STRESS

As you make decisions in the airplane, you are influenced by your mental state and the extent to which you may be under mental stress. We discuss how to recognize stress and how to deal with it.

JUDGMENT

Most accidents are caused by pilot error. Of these, many are a result of poor judgment on the part of the pilot. In this chapter, we discuss what is meant by judgment, how it influences decision making, and how to assess your own tendencies under stress.

COCKPIT RESOURCE MANAGEMENT

Although cockpit resource management is usually thought of as being intrinsic to multi-person crews, there are many aspects that apply to general aviation, in single- and multi-pilot settings. We discuss these and provide exercises that will enable you to assess your own abilities to communicate effectively.

THE FUTURE

In this final chapter, we discuss issues of how airplanes should be designed from a pilot's perspective to help reduce errors. We also introduce some new ideas in training that we feel will improve both the efficiency and effectiveness of training.

In each of these chapters, we take a practical approach to provide you with useful information about human behavior and how people fit into the aviation system. We highlight areas in which humans may be susceptible to making mistakes and provide recommendations for recognizing them. Where possible, we suggest how you can avoid making errors that often lead to problems.

This familiar quotation captures it all:

"Aviation is in itself not inherently dangerous. But to an even greater degree than the sea, it is terribly unforgiving of carelessness, incapacity or neglect."

We owe it to ourselves to be as well prepared as possible for every flight we make.

Why Accidents Happen   1

INTRODUCTION

This book is about the topic commonly called **human factors** and how it applies to general aviation pilots. In this context, human factors deals with errors that pilots make, why they make them, and how they can prevent them. The book discusses the major issues that affect how well pilots fly — issues sometimes within their control, but also often beyond their control. It is a book about human performance and factors that affect it. The following excerpt from an International Civil Aviation Organization publication captures the essence of human factors:

> Human factors is about people: it is about people in their working and living environments, and it is about their relationship with equipment procedures and the environment. Just as importantly, it is about their relationships with other people. Human factors involves the overall performance of human beings within the aviation system; it seeks to optimize people's performance through the systematic application of the human sciences, often integrated within the framework of system engineering. Its twin objectives can be seen as safety and efficiency.

ICAO Circular 227

HUMAN FACTORS

You may well ask what is so important about human factors and the role of the pilot; you may be puzzled why we think that the traditional pilot training topics provide inadequate human factors training. The reason is that more than 80% of the accidents in general aviation today are caused by **pilot error**, not by major malfunctions of aircraft systems. Furthermore, most accidents could have been avoided had the pilots taken the appropriate preventative or corrective actions.

Accident statistics seem to indicate that pilots do not have the appropriate tools to deal with their own roles in the cockpit. They are trained to handle emergencies, such as engine or electrical system failures, but they are not trained to handle the internal and external forces that act on them. In fact, they are often not even aware of the existence of some of these forces.

In this chapter, we set the stage for the rest of the book. We introduce some basic concepts and ask you to complete a short questionnaire that deals with topics covered in the book. We discuss how and why accidents occur and provide accident statistics for general aviation to illustrate the nature of the human factors problem.

DEFINITIONS

In this book, we look at a variety of topics within human factors that have a direct relevance to pilots. Before we begin, however, we need to define what we mean by human factors.

Human factors is the study of how people interact with their environments. In the case of general aviation, it is the study of how pilot performance is influenced by such issues as the design of cockpits, the function of the organs of the body, the effects of emotions, and the interaction and communication with the other participants of the aviation community, such as other crew members and air traffic control personnel.

As we have mentioned, pilot error is given as the primary cause in the majority of accidents. We also need to define what is meant by pilot error.

Pilot error is an action or decision of the pilot that, if not caught and corrected, could contribute to the occurrence of an accident or incident. Inaction or indecision are included in the definition.

We want to emphasize that the term "pilot error" does not imply that all errors are the fault of the pilot. Sometimes external circumstances are the cause, such as poorly designed instruments or controls in the cockpit, or ambiguous regulations or communications. In this book, we discuss both types of errors — those that are the fault of the pilot and those caused by factors external to the pilot. Knowing why pilots make errors is helpful in designing preventive measures, such as creating specialized training programs and redesigning equipment or procedures.

SOME GROUNDWORK

Before we begin our discussion of why accidents happen, we would like you to complete the following short questionnaire. There are 10 questions. On a sheet of paper, write down your answers to each of the questions. Do so without reference to any books, manuals, or friends. Please resist any temptation to formulate the answers just in your head — writing the answers down will help clarify them for you and provide a record you can refer to later in the chapter.

Complete the questionnaire now. Please do not read on until you have completed the questionnaire.

QUESTIONNAIRE

1. You have calculated that a VFR cross-country flight will take one hour. What is the legal minimum requirement for fuel in terms of time?

2. On a VFR flight, are you allowed to fly above an unbroken cloud layer if you meet all cloud separation requirements?

3. Look at the attitude indicators below. In each case, decide which way you would have to roll your airplane to get the wings level with the horizon.

Left   Right   Left   Right

4. Look at the VOR indicator below. Which way would you turn to intercept the inbound radial to the VOR?

Left   Right

5. Assume you are on a normal approach and the runway looks like illustration A below. Later in your flight, you are approaching a runway which looks like illustration B below. How would you judge your approach to runway B (high, good, or low)?

A

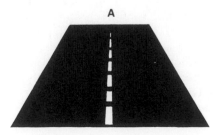

B

6. Read the following numbers:

   1 1 2 1 4 7 4 1 2 2 5

   Now cover them with your hand and write them out from memory. Don't peek!

7. What are the regulations concerning drinking and flying?

8. If you are having marital or relationship problems, will your ability to fly safely be affected?

9. You have an accident while flying in violation of regulations, such as with an expired medical certificate or in clouds with only a VFR rating. In the accident, you cause damage to persons or property. Will your insurance cover the damages? What are the implications of these violations on your personal liability and future insurance coverage?

10. Consider the sentence: "The captain ordered the flight attendant to stop smoking in the cabin." Who was smoking?

Now review your answers in light of the discussion below and tally the points you obtained.

Question 1. You have calculated that a VFR cross-country flight will take one hour. What is the legal minimum requirement for fuel in terms of time?

The legal fuel requirements for cross-country flight depend on whether you are flying IFR or VFR. On an IFR flight plan in the United States, you are required to carry sufficient fuel to fly to your destination, then to your alternate airport, and still have 45 minutes flying time remaining. For VFR flights, such as in our question, the requirements in the United States are: for night flights, enough fuel to reach your destination plus 45 minutes; for day, sufficient fuel to reach your destination plus 30 minutes. Regulations vary slightly from country to country. Give yourself one point if you were correct for both day and night and zero if incorrect or incomplete.

Now answer this question: Have you ever violated these fuel requirements? If you have, subtract one point from your score.

This question tests whether you know a simple fact that you would normally learn in ground school. The reason we asked it is that it is closely related to issues of safety and judgment, as illustrated by the follow-up question. A significant number of accidents are caused by running out of fuel.

Question 2. On a VFR flight, are you allowed to fly above an unbroken cloud layer if you meet all cloud separation requirements?

In the United States, you are allowed to fly VFR over solid clouds. It is not recommended because, if you reach your destination and you are still above the unbroken layer, a descent through the cloud layer would result in a violation of the FARs. Give yourself one point if you said it was legal to fly above a solid layer, two points if you said it was legal but not recommended, and no points if you were incorrect.

Once again, the question tests for a simple fact that is closely related to safety and judgment.

Question 3. Look at the attitude indicators below. In each case, decide which way you would have to roll your airplane to get the wings level with the horizon.

Left   Right   Left   Right

Both attitude indicators show the airplane in a left turn.
You would have to roll right to get the wings level.
Give yourself one point if you answered both correctly.

The point of this question is to draw your attention to
the fact that information can be presented different
ways. It is likely that you had to think about whether
the two indicators showed the same situation. In the
airplane, any hesitation in interpreting the indications
of an instrument could lead to a dangerous situation.

**Question 4.   Look at the VOR indicator below. Which way
would you turn to intercept the inbound radial to the VOR?**

Left   Right

There is no way to answer this question correctly,
because the VOR does not give you any information
about the heading of the airplane. In the diagram
below, both airplanes would have the same VOR
indication. Airplane A would have to turn left to
intercept the radial, while Airplane B would turn right.

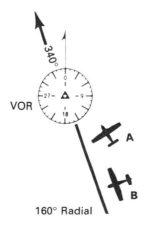

If you said that the question could not be answered, give yourself a point. If you selected either of the choices, subtract a point.

Question 5. **Assume you are on a normal approach and the runway looks like illustration A below. Later in your flight, you are approaching a runway which looks like illustration B below. How would you judge your approach to runway B (high, good, or low)?**

A   **B**

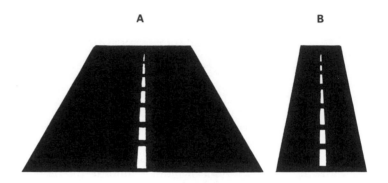

It is impossible to answer this question, because you have to know something about the runway. If the two runways are the same size, you would be high on the approach to runway B. However, if runway B were half the width of A and sloped upward, you could actually be low. The point of the question is to highlight the fact that your eyes cannot always be trusted.

If you chose any one of the alternatives, subtract a point. If you indicated that the question could not be answered, give yourself a point.

Question 6. **Read the following numbers:**

1 1 2 1 4 7 4 1 2 2 5

Now cover them with your hand and write them out from memory.

If you wrote all numbers correctly give yourself a point, and no points if you did not. If we had prefaced the question with the sentence: "The following numbers represent the holidays of New Year, Valentine's Day, Independence Day, and Christmas," would it have been easier to remember them? Most probably, because you would have only had to remember four facts (four holidays) instead of eleven numbers (the number of numerals in the question). Human short-term memory is very limited and can be easily overloaded in the cockpit.

Question 7. What are the regulations concerning drinking and flying?

There are three regulations that deal directly with flying after the consumption of alcohol. The first is that you are not allowed to fly within eight hours of drinking. The second is that you are not allowed to fly under the influence of alcohol. The third is that you cannot fly with .04% by weight or more alcohol in the blood. Give yourself a point if you got all three.

We asked this question because most pilots think that the eight-hour rule is the only one that applies. In reality, the most important is the second: namely, not being allowed to fly under the influence of alcohol because it has no time limit. This raises another question. If you have a reasonably heavy drinking session, say five or six drinks, for how long can the alcohol affect your body? The answer is up to 48 hours! Often the effects linger well beyond your ability to feel them.

Question 8.   If you are having marital or relationship problems, will your ability to fly safely be affected?

If you are having a marital or relationship conflict, it will almost certainly have an adverse affect on your ability to fly safely. It is very important to realize that you cannot separate yourself from what is happening in other parts of your life. Give yourself one point if you were correct; subtract a point if you were wrong.

Question 9.   You have an accident while flying in violation of regulations, such as with an expired medical certificate or in clouds with only a VFR rating. In the accident, you cause damage to persons or property. Will your insurance cover the damages? What are the implications of these violations on your personal liability and future insurance coverage?

The answer to this question varies with the policy and insurer. If you know the terms of the policy that normally covers you, give yourself one point. If you do not, or if you had to read your policy, subtract one point.

We asked this question because it is good judgment to know all the possible ramifications of your decisions. If you ever found yourself in a situation in which you were deciding whether to fly with an expired medical certificate, knowing the full extent of the financial risk could sway your decision.

Question 10.   Consider the sentence: "The captain ordered the flight attendant to stop smoking in the cabin." Who was smoking?

The sentence (and the captain's order) is ambiguous. It is not clear whether the flight attendant was smoking or whether the flight attendant was asked to stop other people (the passengers) from smoking. If you realized that the sentence was ambiguous, or if you indicated that you did not have enough information to respond, give yourself one point.

Sentences like this show how ambiguous language can be. Consider the following: "Take off power." What would you do if you were the co-pilot in an airplane on the takeoff roll and the captain issued this order?

INTERPRETING THE RESULTS

The purpose of these questions is to give you an insight into what this book is about. Some people we have given the test to have complained that the questions are tricky or unfair.

That is not the case. All the questions reflect real issues that affect you as a pilot. In reality, there are a lot of confusing and misleading things that can happen to you in the cockpit.

There are three parts to the interpretation of the results. First, tally your score. If the total is less than 10, there are some deficiencies in the factual information that you use when flying. Although this does not necessarily mean that you are vulnerable to making mistakes, it is definitely easier to make good decisions based on accurate data. If your score was seven or less, you should consider a thorough review of your aeronautical knowledge.

Second, try to recall your feelings as you answered the questions. If you were uneasy because you felt you should know the answers but were not confident that you did, you are at least aware that your knowledge may be deficient. You should be very concerned if you did not score well but felt confident that you knew the answers. In this case, you may be prone to making pilot errors.

Finally, if you had to subtract any points, you may be prone to making errors of judgment. In Question 1, if you have violated the fuel requirements, it is possible that an unexpected change in the weather or wind could lead to fuel starvation and a forced landing. In Question 4, if you think you can tell heading from the VOR reading, you may find yourself lost. In Question 5, if you think that you can judge your glide slope from how the runway appears, you may be in for a nasty surprise when you land at unfamiliar airports. In Question 8, if you think you can leave either personal or business problems behind you when you fly, you are most likely wrong. Problems have a habit of staying with us, even subconsciously, and outside stresses are a major contributor to accidents. In Question 9, the issue is not so much safety of flight as it is what an accident could do to your financial state. You should always know the details of your insurance policy. Of course, you must always fly within the rules established by the FARs.

Each of the questions relates to a chapter in the book. Questions 1 and 2 relate to factual knowledge — information you need to know to be a safe pilot. We devote a chapter to learning and learning environments. Questions 3 and 4 raise issues about the role cockpit design has in how well you fly. We discuss these issues in several chapters. Question 5 highlights how fickle your vision is, and how careful you must be to check for supporting information. We have a chapter on how the eyes and ears gather information which you use when flying. Question 6 was designed to show up some of the limitations of your brain. We have a chapter relating to the brain and how it functions. Question 7 raises issues about how well you know the strengths and weaknesses of your own body. In our chapter on the body, we discuss the factors that can adversely affect your body, as well as indicating things you can do to improve your body's ability to perform. Question 8 deals with the effects of stress on your performance. We have a chapter devoted to emotional stress and ways to minimize its effects. Question 9 raises the issue of judgment, to which we devote an entire chapter. Finally, Question 10 shows how easy it is for communication to fail. This and other cockpit issues are discussed in the chapter on Cockpit Resource Management.

All of the issues raised by the questions are part of the field of human factors, and all relate directly to your ability to fly. When these factors affect you adversely, the chances increase that you may perform badly and have an accident.

ACCIDENT STATISTICS

It is helpful to look at accident statistics for general aviation, because they give you an idea of both the extent and nature of the human factors problem. The statistics we use here come from a variety of sources, most notably the National Transportation Safety Board (NTSB). Their interpretation is sometimes controversial because it is not always clear what is meant by some of the categories. For example, a great number of accidents (about 24% in 1985) fall into the category called weather-related. However, it is possible that pilot error was the primary cause, because the pilot made a conscious decision to continue into known adverse weather conditions.

The good thing about these statistics is that they continue to improve. [Figure 1-1] Almost every year since World War II, both the accident and fatal accident rates have fallen, so that today the accident rate for general aviation is about 8 accidents per 100,000 hours flown, and less than 2 fatal accidents per 100,000 hours flown. In 1987 in the United States,

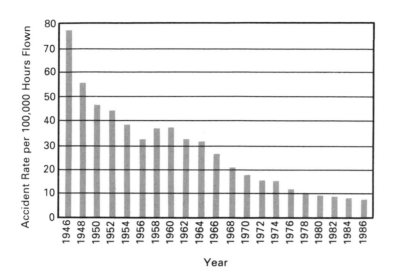

Year

Figure 1-1

there were just over 2,400 general aviation accidents, of which 426 (17.6%) were fatal. In 1985, there were 2,741 accidents of which 498 (18.2%) were fatal.

The bad thing about these statistics is that people continue to crash and be killed. In 1985, the pilot was found to be a "broad cause/factor" in 84% of all accidents, and 90.6% of all fatal accidents. This means that the responsibility for accidents is largely the fault of people, not the machines they fly.

The most common, specific causes of the accidents, in order of frequency, are:

* Loss of directional control
* Poor judgment
* Airspeed not maintained
* Poor preflight planning and decision making
* Clearance not maintained
* Inadvertent stalls
* Poor crosswind handling
* Poor inflight planning and decision making

As you can see, almost all of these are a result of poor pilot performance, not a result of equipment malfunctions.

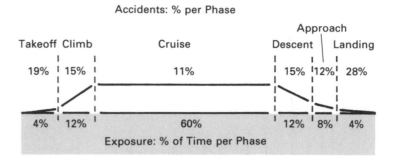

Figure 1-2

Another way to study accident statistics is to determine the phase of flight where accidents are most likely to occur. The 1985 data show that 28% of all accidents occurred in the landing phase, 19% in the takeoff phase, and 12% in the approach. Of the landing accidents, most occurred in flare, touchdown or, most commonly, the rollout where directional control was lost. Figure 1-2 depicts the distribution of accidents by phase of flight and also shows the amount of time spent in each phase.

It is clear that the majority of accidents occur when approaching or leaving airports. This is not surprising, since the workload of the pilot is greatest at these times, which increases the chance of making an error.

A SIMPLE MODEL OF PILOT PERFORMANCE

As you review the statistics in the last section, it is likely that you noticed that there were a variety of types of errors. In this section, we introduce a basis for discussing these various types of errors and lay the groundwork for more detailed explanations in the following chapters.

A very simple model of how the pilot functions is depicted in figure 1-3. Central to the whole model is the pilot, because it is the pilot who controls the system.

The first stage of the model shows information flowing to the pilot. For the most part, this is accomplished through the eyes and ears, although some information can also be gathered through the nose (such as smoke from an electrical fire), the vestibular senses (such as acceleration, deceleration, and turns), and touch (such as something is hot or vibrating).

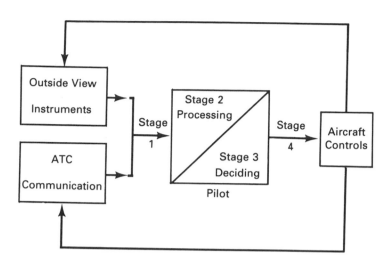

Figure 1-3

Two types of error can occur in this phase — the information itself may be wrong, distorted, or just plain missed; or the correct information is gathered but then incorrectly interpreted by the brain. In both cases, the brain uses incorrect information as it undertakes a decision-making process. Obviously, the less accurate the information used by the brain, the more likely it is to make a poor decision.

We discuss the issues of the quality of information available to the pilot in the chapters on Cockpit Design and The Eyes and Ears. In Cockpit Design, we look at how design issues affect how you interpret flight information and show how this information is often distorted by the way it is presented. In The Eyes and Ears, we pay particular attention to both the physical and psychological aspects of information flow and discuss issues such as physical constraints in vision and visual illusions in flying.

The second stage of the model is the processing of available information by the brain. A number of aspects are important here, such as the ability of the brain to notice available data, its ability to choose between data that are available simultaneously, and the ability to discriminate between relevant and extraneous information. It is also important for the brain to switch rapidly between several tasks that need to be handled at the same time, such as flying the airplane safely and solving some inflight emergency.

Several types of errors can occur in this phase. The brain may be so intent on one task that it ignores readily available information from another source, or the brain may be so bored that it fails to notice changes in the environment. Almost always, information is available simultaneously from a number of sources. Another common error is for pilots to fail to choose the best available data or to check the accuracy of information through cross-checking. Finally, the brain may concentrate so heavily on one task that other tasks are neglected. We devote a chapter to a discussion of how the brain processes available information.

The third stage of the model is the decision-making process. In some ways, this is part of the information-processing phase, but other factors also impinge on the decision-making process. For example, although accurate data may have been gathered by the brain when making a decision, other factors can influence the quality of the decision, such as social or emotional pressures, or lack of oxygen. We discuss some of the factors that can influence decision making in chapters on The Body and Emotional Stress, and we look at the role judgment plays in its own chapter.

The fourth phase of the model deals with the ability of the pilot to implement the decisions made in the third phase. Even if the correct decisions have been made, problems can still occur if there are factors that detract from the ability of the pilot to act. Typical of these are the debilitating effects of fatigue, drugs, or hypoxia to name but a few. The chapters on The Body and Emotional Stress also contain an extensive discussion of a number of physical and emotional issues that affect the ability of pilots to perform effectively.

One other factor that often contributes to errors, particularly in multi-crew situations, is poor communication between the personnel. There have been numerous examples of the solution to an inflight problem being known by one crew member, who is unwilling to share it or whose input is disregarded. We deal with these issues in the chapter on Cockpit Resource Management.

To summarize, errors can be introduced into the decision-making process at each of the four stages. Sometimes the errors are caused by faulty information, sometimes by faulty interpretation of accurate information. Even if accurate information reaches the brain, the brain itself may introduce errors by ignoring useful data, by not noticing data, by failing to focus on all the tasks needing attention, or by not computing the full range of alternatives. Sometimes errors can be introduced into the system by reason of external pressures that encourage the pilot to choose a poor alternative action. And sometimes errors occur even when the correct decision has been made, because the physical or mental state of the pilot detracts from the ability to perform well.

ACCIDENT PROFILES

The NTSB has used all the data it has collected to establish a profile of pilots most likely to have accidents. The profile is useful because it allows you to compare your own current profile. Although you should always be careful, statistics indicate that you should be even more cautious if you fit the accident profile. The pilot most likely to have an accident:

* is between 35 and 39 years old
* has between 100 and 500 hours total time
* is on a personal flight
* is in VMC conditions

Our interpretation of this profile is that there is a period in pilots' careers, between approximately 100 and 500 hours, in which their confidence level exceeds their ability level. Within this envelope, we think that there are two particularly dangerous times. The first occurs around 100 hours, when a pilot has added about 50 hours of flying after receiving the private pilot certificate. The second occurs 50 to 100 hours after earning an instrument rating. In both cases, confidence has grown, but there still has not been much time to gain experience.

Even though this profile helps to understand when pilots, in general, are more likely to have accidents, it does not mean that you are invulnerable if you do not match the profile. The best attitude to have is that you are vulnerable at all times and that you should exercise caution at all times.

CONCLUSION

The information in this chapter can be rather daunting. There are so many ways and so many places that errors can occur, that you may wonder how it is possible for anyone to fly safely. In the next chapters, we discuss all these issues that relate to the process of making good decisions and exercising sound judgment. Our approach is to persuade you to accept the fact that you are fallible and should learn to operate within the constraints that this imposes. If you are aware of the problems of decision making and know what types of errors are likely to occur in different situations, you will be better prepared to recognize and deal with them.

Exercises

Chapter Questions

1. Define Human Factors.

2. Define pilot error.

3. Most airplane accidents are caused by pilot error. Give three examples of pilot error in which it is difficult to blame the pilot for the error. Give three examples in which the pilot is clearly responsible for the error.

4. List five objects that you encounter every day, such as eating utensils, automobiles, computers, etc., that have design features that make them difficult to use. Do the same for objects in which the design enhances their use.

5. Draw a profile of a typical VFR flight in terms of minutes spent on each phase and estimate the percentage of accidents that occur in each phase.

6. Most accidents occur during the landing phase at the end of a flight. Explain why you think this is the case.

7.   What are the five most common causes of accidents in general aviation?

8.   Draw a picture showing the relationship between the pilot and the airplane in terms of control and information feedback.

9.   What two broad categories of errors can occur when a pilot gathers information about the status of a flight?

10. Give the profile of a typical general aviation pilot involved in an accident. Why do you think this type of pilot has accidents?

REFERENCES AND RECOMMENDED READINGS

AOPA Pilot. May 1988.

International Civil Aviation Organization. 1989. *Fundamental Human Factors Concepts.* ICAO Circular #216. Montreal.

International Civil Aviation Organization. 1989. *Flight Crew Training: Cockpit Resource Management (CRM) and Line-Oriented Flight Training (LOFT).* ICAO Circular #217. Montreal.

International Civil Aviation Organization. 1991. *Training of Operations Personnel in Human Factors.* ICAO Circular #227. Montreal.

National Transportation Safety Board. *Annual Review of Accident Data.*

Learning to Fly

2

INTRODUCTION

This chapter focuses on the process of learning how to fly, whether you are a student or an experienced pilot. We discuss a variety of techniques to improve your skills, from more effective ways of learning to deciding where to learn and who to select as an instructor. We also take a close look at some of the new technologies that are being used in pilot training and discuss how you can utilize them most effectively.

The process of learning how to fly is a good example of where human factors can play a major role. On the one hand, when you are a student pilot, you know little about the complex and sophisticated aviation environment and, in the beginning, may be nervous about being high above the ground or surrounded by other aircraft. On the other hand, your instructor is typically a very experienced aviator, who knows aviation intimately.

The successful bridging of this gap does not depend on luck but on the professional relationship you establish with your instructor. As with any relationship, both parties have a contribution to make. From your side, the relationship will be enhanced by having a good attitude and a willingness and ability to learn. From the other side, the instructor's attitude is important because it will greatly influence yours. Your instructor's ability to teach also influences the relationship, both in terms of how well he or she knows what is to be taught and how to teach it. All of these issues are part of the field of human factors.

Learning to fly is different from most other instructional settings because it is much more intimate. Unlike classroom instruction, where there are many students for each instructor, much of flight instruction is one-on-one with your instructor. Fortunately, you almost always can have a say in selecting your instructor. If you are unhappy with the instruction you are receiving, you should make every effort to change to someone with whom you are comfortable.

IMPROVING YOUR ABILITY TO LEARN

We start our discussion of learning how to fly by addressing an area that is your responsibility in the instructor-student relationship: namely, your ability to learn. Certainly, there is nothing we can do about your innate abilities, but we can introduce techniques to help you improve how you learn. This is helpful because the volume of information you have to assimilate to be a safe, successful pilot is substantial. There is no doubt that adopting good study methods can have positive effects on both your efficiency of learning and your long-term retention of the material.

At the same time, we want to dispel any ideas you may have that there are "wonder" techniques that can make learning easy. There are NO such things. Learning can never just happen by itself. It always requires effort. So, save your money. Do not buy audio tapes that teach while you sleep — there is no evidence at all that these are effective. Even professionally produced videotapes will not teach you anything unless you expend the effort to learn. If you use them properly, they are very helpful, but they have no magical properties that will make you "know" the information without effort.

The best way to learn is by being an active participant in the learning process. The more you process the information to be learned, the better you will retain it. Moreover, the more meaningful the processing, the better the learning will be. We now introduce a variety of activities that enhance learning and retention. We recommend that you try each of them to see which suits you and your learning style best.

MAKE NOTES

One of the simplest techniques for improving learning is making notes, not just taking notes. What we mean by this is that for each topic you study, you should make a set of notes for yourself using all available sources of information. This includes your notes from the instructor's lectures, textbooks, and other supplementary materials. Figure 2-1 is an example of how to organize your notes from several sources. Borrow-

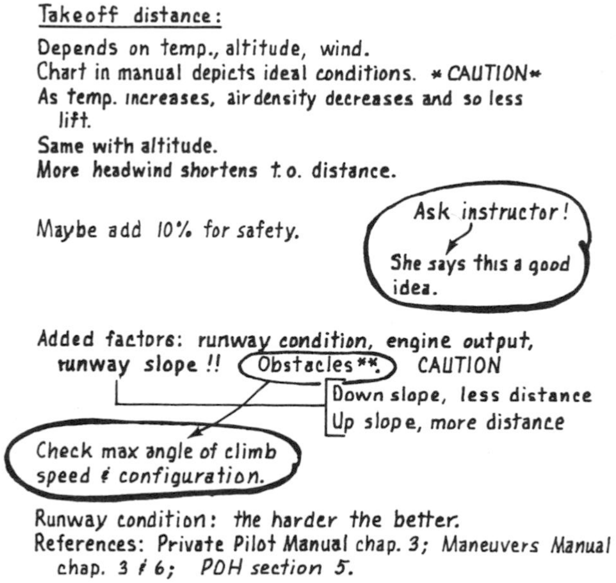

Takeoff distance:
Depends on temp., altitude, wind.
Chart in manual depicts ideal conditions. *CAUTION*
As temp. increases, air density decreases and so less
   lift.
Same with altitude.
More headwind shortens t.o. distance.

Maybe add 10% for safety.

Ask instructor!
She says this a good
idea.

Added factors: runway condition, engine output,
   runway slope !! Obstacles** CAUTION
                    Down slope, less distance
                    Up slope, more distance
Check max angle of climb
speed & configuration.

Runway condition: the harder the better.
References: Private Pilot Manual chap. 3; Maneuvers Manual
   chap. 3 & 6; POH section 5.

Figure 2-1

ing someone else's notes can be helpful in seeing how they have organized the material, but it does little for your own learning.

Typically, you make these notes after all the instruction on the topic is completed. It can be done on paper or on a computer, using one of the many word processing or hypertext programs now available. As you prepare these notes, write down any questions you have about issues you do not fully understand. Do not just make a mental note to ask your instructor, because having a written record ensures you will not forget the question. When you get your answer, write it next to the question, so you will not forget it later. You can do the same thing with this book. There is ample space at the edge of each page to write notes to yourself and questions for your instructor.

Organizing the information into note form helps you process it, which leads to better comprehension and retention. It also means that you have the information organized in a way that suits you. Furthermore, you can keep these notes and refer to them later whenever you need to refresh your mind.

ANSWER QUESTIONS

A familiar technique for processing information is answering questions. You have done this in class, homework, discussions with your instructor and fellow students, and tests. Answering questions not only helps you learn, it also gives you an indication of how well you know the material you are studying.

There is a way of making this technique even more useful. Once you have answered a question and checked your response, ask yourself why it was asked and rank it on a scale of zero to three as to its relevance, with zero being not relevant and three being highly relevant. This helps you think more deeply about the material. You can then discuss your ratings with your instructor to see whether you were correct.

A word of caution is necessary here. Many people study for their FAA Written Examinations by buying a study guide that contains questions that are either directly from the examinations or that are very similar to them. This technique will certainly help you pass the examination. However, there are several important drawbacks you need to consider. First, studying purely for a test almost certainly will result in you remembering the material for only a short time. Much of what you learn will be soon forgotten. Second, rote memorization, which this technique uses, fails to ensure that you understand the material in a useful and meaningful way. If a set of circumstances arises that is different from what you have learned, you probably will not have the skills or information to deal with it. And, third, tests cannot cover all aspects of flying and you may miss an important point altogether simply because the study guide did not cover it.

What we are saying is that you have an important choice to make as you learn to fly. You can be satisfied merely to pass your examinations and obtain your certificates and ratings, which will certainly allow you to fly legally; or you can choose to approach flying with a professional attitude, whether or not you fly for a living. We believe that opting for the former will be your first error in judgment, and you will knowingly be depriving yourself of information and skills that could one day save your life. Choosing the second approach implies that you will not be satisfied by knowing enough to pass; you also desire to know why things happen and how things work. We hope you make the professional choice.

PREPARE QUESTIONS

We have just discussed how answering questions about the material you are studying is a helpful learning tool. However, it is even more helpful to prepare questions about it. This forces you to make decisions about the relevance of the content and requires you to be able to come up with the correct answer. For example, if you are studying aircraft performance and decide to prepare some questions about takeoff distance in different situations, you will first have to decide what affects takeoff roll. Then, you will have to prepare a question with appropriate conditions, such as pressure altitude, temperature, weight, wind, and runway surface. Finally, you will have to solve the problem yourself.

You can take the process two steps further. First, you can ask your instructor to check your answer, which is helpful to your understanding if you have made an error. And, second, you can give the question to your friends, which will help them if they do not know the answer and force you to explain how you obtained the correct answer. Each of these activities helps you process the material which, in turn, leads to enhanced learning and retention.

DRAW A CONCEPT MAP

People vary considerably as to whether they prefer to learn from text or from more graphic sources. If you are more pictorially orientated, this technique may appeal to you. A

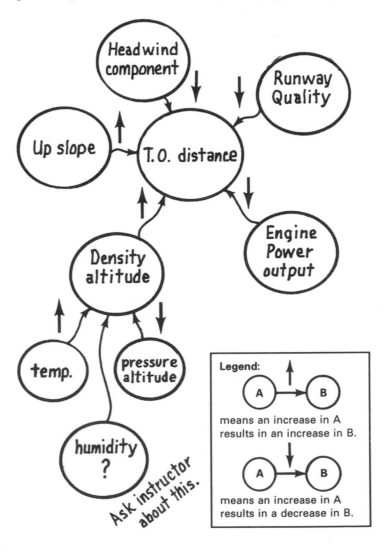

Figure 2-2

concept map is a graphic depiction of the relationships between ideas, information, or concepts. A typical concept map has a number of associated concepts, usually shown in circles, together with the relationships between them, shown by annotated lines.

Figure 2-2 is an example of a concept map that deals with calculating the amount of runway needed to take off. At the center is the focus of the map, namely takeoff distance, and surrounding it are the things that influence this distance. We have drawn the map with two types of connectors to the center: those that lengthen or those that shorten the takeoff distance. For instance, one of the factors that influences takeoff distance is the quality of the runway. The circle representing this is connected to the center with a "shorten" relationship. This means that as the quality of the runway increases, the distance needed to take off becomes shorter. Examine the map carefully and decide whether you agree with it.

When you create your own concept map, the process of deciding what should be included and what the interrelationships are forces you to be explicit about the knowledge in your head. Frequently, we think we know something well, but when we have to draw it or explain it, it becomes apparent that our knowledge is faulty.

It is important to remember that there is no single, correct concept map. There are many ways to represent the relationships. What is nice about the technique is that you can create a personal map that is meaningful to you and the way you think. As an additional benefit, the map can be evaluated by your instructor or colleagues to assess whether your knowledge is accurate. If you looked carefully at the concept map in figure 2-2, you may have noticed that the relationship between pressure altitude and density altitude is wrong. It incorrectly shows that increasing pressure altitude results in a lower density altitude. As you know, the opposite is true; as pressure altitude increases, density altitude increases.

TEACH A TOPIC

One of the best ways of finding out whether you know a topic is to teach it. You can do this either informally or formally. On the informal side, you can offer to explain the material to a class or a friend, or you can prepare some questions on the material and be prepared to evaluate a friend's answers and to explain why they were wrong. A formal approach would be to make a presentation of the material to a class.

One caution is in order. In a teaching situation, you have a great responsibility to ensure that the content is accurate. Even if you are just trying to explain a concept to a friend, it is important that you are certain of the accuracy and not pass along your own misconceptions. In a classroom setting, students may assume that you have prepared the material thoroughly and accurately, and would be even more likely to believe what you were saying. If you do decide to make a formal presentation, you may want to have your instructor review your presentation beforehand for accuracy.

Teaching a topic forces you to organize the subject into a logical order and to establish the appropriate relationships between the various pieces of information. This type of processing is an excellent way of ensuring that you know the material and it results in good retention.

ASK "WHY?" OR "WHY NOT?"

Educational research consistently shows that if you can relate new information to old and can see how new material is relevant to your overall learning goals, then you have a much greater likelihood of remembering the new data. A good way of doing this is to ask a lot of questions about why you are doing something, why it is important, and why it is true. It is also often helpful to ask why you are not or should not be doing something.

Typical of these questions are the following. "Why do I have to obtain a weather briefing for a local flight?" or "Why does an aft CG make the plane more unstable?" or "What benefit is there from learning to fly a chandelle?" or "Why should I always fly a traffic pattern when approaching to land?" Each of these questions helps you understand what you are learning and why.

This technique is also helpful for ensuring that you do not take things for granted. For example, if your instructor is teaching you about stalls and tells you that the stall warning horn comes on in time for you to prevent a full stall, you may want to ask how the stall warning indicator works, whether there are any exceptions to this rule, and whether an electrical failure would change this situation.

Similarly, if your instructor has always told you to complete a weather briefing before all flights and, one day, tells you to skip it, it would be reasonable to ask why the briefing is no longer necessary. This is a good habit to form, because it is situations like this, where routine procedures are changed unexpectedly, that increase the chances of an accident. When you ignore a good procedure because of pressures of time, you can find yourself in double jeopardy — the procedure is not done at a time when it is most needed and, if something goes wrong, you may be too stressed to handle an ensuing emergency.

The process of asking "why" and "why not" questions often encourages an ongoing discussion of what you are doing and what you are learning. Such an exchange of ideas improves learning and understanding, as well as creates a rapport with your instructor and colleagues.

MENTAL PREPARATION

At one time or another, everyone gets rushed. A meeting or traffic jam may have made you late for a sporting event, to catch a flight, or for a cross-country flight to an important customer. In such situations, it often happens that you forget something, such as the tickets for the game or plane, or perhaps to check the fuel before the flight. In the rush, it is easy for you to let items slip from memory, which is why we recommend the use of checklists in the airplane.

A similar situation occurs in training. For example, if you are learning to fly and have a lesson on flying holding patterns, it will detract from your learning if you jump into the airplane and immediately start practicing. It is far better to fly a pattern or two in your head before you start, imagining the type of entry you will make, and the headings you will fly in various wind conditions. This brings into your mind the necessary knowledge and procedures you will need to perform this maneuver. This technique also refreshes your memory on all the things you need to remember.

When you have finished practicing, it is again helpful to mentally replay what you did. This will help you understand what you did, and will reinforce what you learned. Although the research on this technique is not as clear as it is for the others, we have found it to be useful.

STRUCTURED PRACTICE

You may have noticed that most of the activities mentioned so far do not deal at all with actually flying the airplane. Most deal with learning facts, concepts, and processes. In fact, if you make a list of things that a safe pilot has to do, you will see why we place so much emphasis on the mental skills. The list would include:

* Remembering factual information
* Remembering and using proper procedures
* Controlling the aircraft
* Communicating clearly
* Solving known problems
* Solving new problems
* Exercising good judgment
* Improving performance

It is helpful to note that all but one of these are mental skills. Controlling the aircraft is the only physical or psychomotor skill. Safety in flying is actually dependent more on good thinking and processing skills than it is on how well you manipulate the controls.

Nevertheless, this does not mean that the best way to improve your flying is by sitting on the ground with a book. The best form of active learning occurs when you perform. This does not mean that you will learn everything by simply flying. It is quite possible to fly and learn very little. If you want to put your time in the air to good use from a learning perspective, it is essential to have a plan before you go. Know where you are going, what to look for, and what you are going to do. Your two best sources for this information are your instructor and a printed syllabus which you can obtain from either your instructor or a reputable publisher.

For example, if you are going to practice holding patterns, you should know before you takeoff which holding fix to use, which entries to practice, and the orientation and direction of the holding patterns. All of this can be established in the preflight briefing, which also allows you to clarify any questions you may have. Imposing this type of structure facilitates your learning and makes your training budget go further.

Furthermore, we recommend that you consider using simulators and ground trainers as much as possible for those activities that do not need a real airplane. Flying is fun but, in some cases, the airplane is not a good learning environment, and you can save a great deal of time and money by using proper training devices. If you are practicing approaches in a simulator, for example, your instructor can reset you to an appropriate starting position after each approach, rather than having you fly there as you would in a real airplane. This can often more than double your efficiency and save you money that can then be spent on flying for pleasure.

HANGAR TALK

Finally, "hangar talk" can be very helpful. By this we mean that listening to the stories of experienced pilots can be very instructional, both for content and motivation. It is useful to remember that almost always these stories tell of how the pilot got out of a situation that a cautious pilot would never have gotten into in the first place. You should try to find out what caused the situation or problem and how the pilot arrived at the final successful solution. The one potential problem with these kinds of stories is that they have a tendency to glamorize or romanticize risk-taking, and are sometimes used more for inflating the teller's ego than providing a learning experience. Nevertheless, they can be useful sources of information.

All the techniques and activities we have discussed improve your ability to learn by making you an active participant in the learning process. The only time we do not encourage "learning by doing" is when the learning situation is dangerous, such as flying a plane that exceeds maximum allowable weight and has a center of gravity out of the acceptable envelope, or practicing stalls near the ground. In such cases, you should either learn from the experiences of others or practice in an environment, such as a simulator, that mimics reality in the essential features, but is still safe.

CHOOSING WHERE TO FLY

In the previous section, we discussed techniques to improve how you learn. In this section, we provide a variety of background information to assist you in making decisions as to where you should learn to fly, where you should upgrade your current qualifications, or where you should seek recurrency. The factors on which you should base your decision are the quality and content of the academic or ground school syllabus, the quality and content of the flight instruction, and the available facilities.

General aviation pilot training around the world has not changed much over the years, although there are some exceptions both to the content and methods of instruction. Traditional training is divided into two parts: ground school and flight training. In ground school, you are expected to assimilate the knowledge of flying, the rules and regulations, the principles of flight, how engines work, how to navigate, and so on. In the flight training portion of the syllabus, you learn to manipulate the controls of the aircraft and to interpret the information provided by instruments, air traffic control, outside visual references, as well as from other sources. Each part of the syllabus culminates in a test, with the ground school using a written form and the flight training a practical test in the airplane.

GROUND SCHOOL TRAINING

Ground school is the part of learning to fly that deals mainly with mental activities. It is where you learn the factual information about flying, such as the regulations and symbols on charts, as well as the procedures for such things as landing at a nontowered airport or filing a flight plan. We divide our discussion of how to evaluate the ground school component of aviation training into two parts: namely, the **content**, or what is taught, and the **instruction**, or how it is taught.

GROUND SCHOOL CONTENT

Wherever you are in the world, ground schools cover the same three major topics: the airplane, the environment, and the regulations.

When studying the airplane, you learn about how an airplane flies, how the engine and electrical systems work, and how to calculate performance limitations. The syllabus dealing with the environment covers the weather and its effect on flight. You also learn how to obtain weather-related information and how to interpret it. In regulations, you learn about the rules that govern you as you move through the airspace system and the procedures that enable you to fly safely.

Although these broad topics are the same worldwide, the specific content varies considerably and should form one of the bases of your decision. Some ground schools are designed to provide you with sufficient information to pass the written test for the certificate or rating, often in the shortest possible time. Others teach substantially more, believing the information to be necessary for being a safe pilot. We believe the latter approach to be the preferable. One of the problems of ground schools that teach to the test is that a passing grade is usually about 70%. This means that if you know most of the material well, you can overlook or skip the rest without really affecting your ability to pass. What can happen, of course, is that you might overlook content needed to fly safely.

Another trend in aviation that has come about with the increased automation of aircraft and systems is to learn only what is needed to operate the airplane. This often becomes translated into "If this light comes on, then press this button." If you do not know why the light comes on in the first place and what effect pressing the button has on the various systems, then there is little chance that you have the information to decide what to do if pressing the button does not solve the problem.

Recently, some schools have recognized that the pilot is an integral part of the aviation system and have introduced human factors as a fourth focus of ground school instruction. In Chapter 10, The Future, we discuss some of these changes and their impact on training.

Your decision about the type of ground school to select, should include a detailed inquiry about the syllabus. If possible, obtain a copy of it that includes content, time spent on each topic, and accepted standards of performance. This will allow you to determine what the ground school teaches and the depth of coverage of the content.

Our recommendation is to be very curious about everything you learn. Try to understand how systems work and why things go wrong. Know as much as you can about everything that has an impact on your flying. Your choice of a ground school should meet these needs.

GROUND SCHOOL INSTRUCTION

The second component of choosing a ground school is the quality of instruction. This is more difficult to ascertain because there are no readily available indices of instructional quality. Nevertheless, your ground school instructor is very important to your education because, as we mentioned above, most of what you need to learn is mental.

There are several approaches to determining the quality of the academic instruction. First, you can interview the instructors. Second, you can interview the students. And, third, you can look at the available instructional support.

INTERVIEWING GROUND INSTRUCTORS

If at all possible, you should take the opportunity to spend time with the ground instructors to find out about their teaching style and methods, the currency of their knowledge, and how well you relate to them. You should do this after reviewing the syllabus so you are well prepared for the discussion. You may also want to read the Teaching Flying section later in this chapter to become familiar with some of the techniques of good instruction.

It is always helpful to have some of your questions prepared beforehand so you won't forget anything important. The following questions are typical of those you may ask. We annotate each with the reason why you ask it.

"Do you follow the published syllabus?"

> This establishes whether the syllabus information you have is accurate. If there is a discrepancy, you should find out what the differences are and why they have been implemented.

"Do you provide written notes of your lectures?"

> This assumes that classes are taught formally in a classroom. If instruction is less formal, the question should be rephrased appropriately. The reason for asking this is that you cannot pay as much attention to the content if you have to write down what the instructor is saying. Having written notes also implies good organization.

"Do you have a 'need-to-know' approach or do you cover topics in depth?"

> The answer to this question should help you decide whether the instruction will help you answer the "why" questions about flying, such as "Why is there an acceleration error in the magnetic compass?" or "Why can the carburetor ice up in summer?"

"How do you deal with slow students?"

> This is an important question if the instruction is to be given in a classroom setting. It is always a difficult line for an instructor to draw. On one hand, good students want to move forward quickly and become bored if they have to wait; on the other hand, weak students get lost if instruction proceeds too quickly. A balance must be found, which frequently involves some explanation in class to meet the needs of the slower students, followed by extra time spent after class, either individually or in a small group.

"Do you encourage students to ask questions in class?"

> This is similar to the last question and can reveal whether the instructor is under time pressure. If the instructor is reluctant to answer too many questions, you may have to find someone else to help you.

"How frequently do you give quizzes and tests?"

> Frequent quizzes and tests are helpful both to the instructor and to you. It is better to find out what you do not know during instruction than during the certification examination. Quizzes and tests allow you to seek clarification facts from the instructor and to correct potential misunderstandings.

"Do you give detailed feedback on quizzes and tests?"

Even though knowing whether you answered a question correctly is helpful information, it is far more useful to have an explanation of why your answer is incorrect. In this way, the quiz or test becomes an integral part of the learning environment instead of an obstacle that generates nervousness.

"Where do you get the questions for your quizzes and tests?"

This is a difficult question. If the questions come only from a bank of questions similar to or the same as the certification examination, you need to explore whether the instructor's aim is purely to help you pass the test. If the instructor creates the questions, you should find out whether they are all multiple-choice or whether they are short answer. Short-answer questions are usually a better test of your knowledge than multiple-choice. However, they are much more demanding on the instructor in terms of time to grade the responses.

"To what extent is human factors covered in the classes?"

This enables you to find out whether issues of human factors are included in the syllabus and whether they are regarded as being important.

These questions, or similar ones that you prepare, will give you a good sense of what to expect in the ground school instruction. Spending time like this will also give you a good idea of whether you would respect and like the instructor.

INTERVIEWING STUDENTS

It is also a good idea to meet students who are already enrolled in the program you are considering and to ask them questions similar to the ones above. It can happen that the instructor and students will have slightly different perspectives, which is helpful to find out. We have always found that students are perceptive and can offer valuable insights into issues such as the quality of instruction and the relevance of content.

INSTRUCTIONAL SUPPORT

The amount and type of instructional support varies greatly from one flight training operation to another. Many ground schools continue to use the traditional instructional aids such as slides, videotapes, overhead projectors, and a variety of models (system diagrams and cutouts). These are all useful if used properly and will facilitate your learning. Some of these, such as videotapes, are particularly useful if they can be checked out and viewed on your own time. If you do this, view the tape with a friend so you can discuss issues that you do not understand.

Some ground schools are also beginning to use computers and their related peripherals, such as videodisc technology, to enhance instruction. All of this computer-related, instructional technology is usually called **computer-based training** (CBT). CBT can comprise a single computer or a network of terminals that can interact. When a videodisc is used for instruction, it is often called **interactive video**, and it can be used with little or substantial computer involvement.

CBT has different forms, most of which are useful in training. Some of them, such as tutorial instruction and testing, apply more to ground school topics than to flight training. Tutorial instruction is typified by the program presenting information on a topic, asking you questions, and giving feedback as to the correctness of your response. Computer-based testing enables you to answer large numbers of questions, which are usually multiple-choice, on many topics and receive feedback on whether you are correct. Both of these CBT methodologies can be used without your instructor being present and allow you to proceed at your own pace. Some schools will also allow you to check out computer programs to use at home if you have a computer there.

Good supplementary instructional materials can enhance your learning. Their availability should be another factor in your decision about which ground school to choose.

FLIGHT TRAINING

We now turn to the second part of your quest to find a place for learning to fly: namely, assessing the quality of the flight portion of the instruction. This is where you learn the skills to take off, fly, and land an airplane safely. As we have mentioned several times before, flying is predominantly a mental activity, but it certainly does require well developed physical or psychomotor skills. The two most important of these are the physical skill of controlling the airplane through the use of the flight controls and the psychomotor skill of observing and analyzing the environment both inside and outside the cockpit. Typically, the first is easier to learn than the second. We divide our discussion into three parts: namely, the facilities, the content, and the instruction.

FLIGHT FACILITIES

The most important element to look at during your review of the flight school is the airplanes used for training. You should check the equipment that they have, what they look like inside and out, and examine the maintenance records. We realize this can be a difficult task if you are new to flying, but you should try to find an experienced pilot or mechanic not connected with the organization to help you.

The aircraft should be properly equipped for everything you will need for learning. This will be different, for example, if you are a student pilot studying for a Private Pilot Certificate, or if you are studying for an Instrument Rating.

How the airplanes look can give you a good sense of the care with which they are maintained. Clean airplanes usually indicate that maintenance is a high priority, which is important because your safety depends on it.

The examination of the maintenance records will tell you how frequently the airplane is in the shop being repaired and the type of problems it has experienced. You should also inquire whether the organization uses progressive maintenance, which means that it is ongoing, or whether all maintenance is done at the 100-hour or the annual inspection. Both of these approaches are fine, but they can have an impact on the availability of the airplane. The final thing to check is the availability. Some organizations have airplanes dedicated to instruction, others rent airplanes to both students and regular customers. If the airplane is frequently rented for out-of-town flights, you may not be able to fly when it is convenient for you.

The other element of the facility worth investigating is where you prepare for a flight. Typically, these facilities should include good access to the services provided by the flight service station, access to weather information, and a place for you to lay out your maps and charts as you prepare your flight. There should also be a comfortable, quiet area for pre- and post-flight briefings.

FLIGHT CONTENT

Assessing the content of the flight portion of learning to fly is usually easier than the ground school, because it is better defined. Once again, you should request a syllabus to determine the extent to which the overall instruction has been planned. Some syllabi allocate hours to each lesson, others only the sequence of lessons. Each lesson should have clearly defined completion standards. For example, a lesson on steep turns may have the following completion standards: maintain between 40° and 50° of bank, remain within 100 feet of the assigned altitude, roll out within 10° of the assigned heading, and maintain smooth, coordinated flight throughout the maneuver.

There are some content areas that are becoming increasingly important as the airspace increases in complexity. These include operations into and out of controlled airspace, such as TCAs; use of advanced avionics; and familiarity with all available ATC services. If the instruction is to be given at an airport with a tower, you should check the syllabus to make sure you will receive additional training at nontowered airports. If the instruction is to be given at a nontowered airport, the converse is true.

Another content area that is being adopted, albeit slowly, by some flight schools is Cockpit Resource Management (CRM). This is similar in concept to the CRM used by the major air carriers, but differs in that it is tailored for general aviation. Although you may wonder about the relevance of CRM in general aviation, with its large fleet of airplanes that require only one crew member, we think that there are benefits worth considering.

First, many pilots aspire to fly larger airplanes that require more than one crew member. If you learn to fly as a single pilot, the habit of doing so is difficult to change when there are other crew members. Effective communication does not come easily when you have had to rely on yourself for your whole career. This is particularly true when pointing out an error a colleague has made, such as telling the captain that fuel is low or that an incorrect frequency has been dialed into a radio.

Second, it is very common to have another pilot in the right seat even though a second crew member is not required. Sometimes this is your instructor, sometimes a pilot friend. Knowing how to communicate effectively can help prevent problems arising from both people thinking that the other has responsibility for an action. In addition, effective communication with ATC is also a benefit, particularly if there are problems with a flight. We deal with the topic of cockpit resource management in a chapter of its own later in this book.

FLIGHT INSTRUCTION

The third component of choosing where to take flight instruction is the quality of your instructor. As with ground schools, this is difficult to determine because of the lack of readily available indices of quality. Nevertheless, this quality is very important to your education because you will have a strong tendency to model yourself after your instructor. The manner in which your instructor conducts himself or herself is likely to impact your flying skills for the rest of your life.

INTERVIEWING FLIGHT INSTRUCTORS

As with the ground school, you should spend some time interviewing several flight instructors to determine their teaching styles and methods. You should do this after reviewing the syllabus so you are well prepared for the discussion. Your personal relationship with your flight instructor is even more important than with your ground instructor, because the instruction is one-on-one and takes place in a dynamic, sometimes frightening, environment. Your aim should be to select a flight instructor you trust based on his or her professionalism.

Once again, it is always helpful to have some of your questions prepared beforehand so you won't forget anything important. The following questions are typical of those you may ask. We annotate each with the reason why you ask it.

"Do you follow the published syllabus?"

> This establishes whether the syllabus information you have is accurate. If there is a discrepancy, you should find out what the differences are and why they have been implemented.

"Do you provide a thorough briefing before every flight?"

> It is highly beneficial to your learning to understand precisely what you will be doing during each flight. The preflight briefing is the time to clarify any misconceptions you may have about the upcoming lesson. Without such a briefing, you may spend time in the air without learning much.

"Do you provide a thorough debriefing after the flight?"

> This serves a similar purpose to the preflight briefing. It allows you to sort out any problems you may have had and to consolidate what you learned. This type of verbal review will help ensure your retention of what you did.

"Do you have a 'need-to-know' approach or do you cover topics in depth?"

> The answer to this question should help you decide whether the instructor will help you answer the "why" questions about flying, such as "Why did the light come on?" or "What happened to the systems when I pressed the button in response?"

"How do you deal with slow students?"

> This is an important question, especially if the syllabus allocates a particular amount of time to each lesson. You should find out whether the instructor is required

to follow the published times and, if not, how the additional time is handled in terms of scheduling and cost.

"What is your attitude toward the use of simulators and ground trainers?"

This question gives you an insight into the instructor's philosophy on instruction. We discuss simulator instruction later in the chapter.

"Do you mind if I ask lots of questions?"

Some instructors like their students to ask lots of questions because it helps their instruction. Others prefer students to listen intently without asking questions. These are different styles and you should choose the one that suits you best.

"What happens if you are not available for a lesson? Who will be my instructor?"

It is useful information to know who the backup instructor is in case yours becomes ill or leaves. You should spend time with the backup instructor as well.

"How do you deal with issues of human factors?"

This enables you to find out whether issues of human factors are important to the instructor and, if so, how they are to be worked into the instruction.

These questions, or similar ones that you prepare, will give you a good sense of what to expect from your instruction and whether you will respect and like the flight instructor. It is helpful to remember that, as the paying customer, you have a say as to who your instructor will be, and you do not have to accept whomever is assigned. It is worth spending the time to ensure you have the right instructor.

INTERVIEWING STUDENTS

As with the ground school, it is a good idea to talk with some of the students who are already enrolled in the program and find out what they think of the flight instructors and instruction. In addition to asking them questions similar to the ones above, you should try to find out more about the

techniques for both ground school and flight instruction. If you are already a flight instructor, we hope that you can use the information to refresh skills that you may have forgotten. If you are aspiring to become a flight instructor, the information will form a good foundation on which to build your skills. We also have a short section on using simulators effectively.

PRINCIPLES OF GROUND SCHOOL INSTRUCTION

This section covers some of the basic principles that contribute to good instruction in the classroom. Good instruction will motivate students and help them to learn the material better. It also can make teaching ground school more pleasurable. The principles we discuss actually apply to all instruction, not only what occurs in the aviation classroom.

First, learning is improved if the relevance of the material is made clear. That is, it is easier to assimilate information in context. This implies that you should always introduce new material with a discussion of how and why it fits into the overall scheme of things.

Second, learning is improved if new material builds on existing knowledge. That is, if you are going to introduce a new concept, your students will learn it better if they already know what are called the subordinate concepts. For example, you cannot effectively teach the concept of stalls, if your students do not understand the subordinate concepts of lift and angle of attack. This has two implications: you should structure the course content to ensure this building block approach, and you should check that the students know the subordinate material before introducing new information. This can be accomplished through written or verbal tests and quizzes.

Third, students learn better if they are active participants in the learning process than if they are passive. We have discussed this at some length earlier in the chapter, but it implies that you should create situations that encourage your students to be active. One easy way to do this is to ask your students lots of "why" and "why not" questions.

Fourth, you should identify and correct student errors as early as possible and not allow them to become habits. This implies that you provide as much feedback as you can in the early parts of instruction. In particular, you must structure feedback in such a way to provide maximum help. It must be corrective. It does not help very much if you tell your students that they are wrong without telling them why. Feedback must be respectful. It is damaging to your students' attitude and morale if you verbally abuse them by shouting or swearing at them. Students rarely make mistakes on purpose, so this type of behavior is particularly inappropriate. And feedback should be discriminating. That is, if a student makes an error due to a misunderstanding, then you should clarify the misunderstanding.

If a student in the United States, for example, says that the VFR minimums are 1 statute mile visibility, 500 feet below clouds, 1,000 feet above them, and 2,000 horizontal separation, you should respond that these are correct only under certain conditions (day, outside controlled airspace, and between 1,200 feet AGL and 10,000 feet MSL). If the student cannot provide the VFR minimums for the other conditions, then you should provide remedial instruction that explains why different conditions require different minimums and what these are. It is not helpful merely to say that the student is only partly correct or even wrong.

And, fifth, you should match the media you use for instruction to what is being taught. For example, trying to explain how a TCA works without showing students a picture of one is very difficult and will likely lead to confusion. Similarly, teaching meteorology without photographs or drawings of clouds, weather systems, weather charts, and so on, will make the topic much more difficult. In this area, computers are beginning to provide excellent instruction for some topics, but care needs to be taken that they are not just expensive, automated page turners. They should enhance current instruction, not just be a substitute for it.

In addition to the general principles of instruction, there are some practices that are specific to flying that can make a difference. For example, your attitude makes a significant difference in formulating the attitudes of your students. If you rush through the academic material, there is no doubt that the students will be left with the impression that the

cognitive or intellectual content of flying is of little importance. Similarly, if you always indicate that you think that
the time spent in ground school is wasted and would be
more profitably spent in the air, you leave the same impression. The students will then place their emphasis on learning
in the airplane. In some senses, this is natural because their
goal is to fly. However, to teach merely the mechanics of
flying and not the associated knowledge and mental skills is
inappropriate.

Another practice that makes a significant difference is setting
a good example. If your actions are in conflict with what is
taught, students are more likely to follow what you do. For
example, if you teach that planning is an important part of
safe flying, but then show little evidence of it, students will
tend to minimize its importance. As they say, "Actions speak
more loudly than words."

Ground school instruction is very important for safe flying,
and will become more so in the future as aircraft and their
systems increase in complexity and automation, and as the
airspace becomes more regulated. Flying is mainly mental
and it will become increasingly more so. We hope that you
will rethink your syllabi and teaching methods on an
ongoing basis so your students come to understand that there
is more to flying than stick and rudder.

PRINCIPLES OF FLIGHT INSTRUCTION

The most exciting part of learning to fly is the time spent in
the airplane. After all, this is why we learn — so we can fly.
As an instructor, you have control over what occurs in a
flight lesson. The following principles should assist you in
maximizing the effectiveness of your student's time in the
air. The basic principles are: be professional by setting a good
example and be prepared.

PROFESSIONALISM

Of all the people who influence a person learning to fly, the
flight instructor has the greatest impact. It is imperative,
then, that you set an example that is conducive to your
student adopting a professional approach to flying.

There are several aspects to being professional we want to mention: keep your knowledge current, be punctual, alert your student in timely fashion to changes in schedule, be prepared (more on this in the next section), be patient, and exercise good judgment. These are all attributes of a professional that leave a good impression on your student. Remember, your student is buying your time and deserves to obtain value for money spent.

The final aspect of a professional approach is how you conduct yourself in flight. The accepted way of teaching a new maneuver is to brief it thoroughly before the flight, then demonstrate it yourself in flight while you talk to the student about what you are doing. Then, let the student try it, perhaps in parts first. As the student learns the maneuver, it is good practice to provide immediate feedback so your student does not form bad habits. As the student improves, the feedback can be reduced and delayed until the maneuver is completed.

Finally, never lose your temper, swear, or shout at your student, even if you feel provoked. It is a better practice to deal with the situation in a civilized manner, even if it means you terminate a flight so you can have a discussion on the ground.

PREPARATION

The principle of flight instruction that has the greatest impact on the student is how well you are prepared for each flight. Your preparation should include matching your student's current performance standard with what is to be taught and then preparing the lesson.

Each lesson should include a thorough briefing prior to the flight. This includes the purpose of the lesson, how it fits into the syllabus, the components of what is to be accomplished, what the student should be doing to accomplish the task, and the standards to which the student is expected to fly. You should also try to manage the lesson in such a way that the time is used efficiently and effectively. If you are going to repeat a maneuver several times, such as takeoffs and landings or approaches, plan to use the time between completing one and flying back to the beginning of the next in a productive way. For instance, you could take the controls so your student is able to relax and listen to your feedback.

You should also plan time for a thorough debriefing after the flight. This enables the student to consolidate what was learned throughout the flight, and it provides an opportunity to ask questions to clarify any misconceptions.

PRINCIPLES OF SIMULATOR INSTRUCTION

As a flight instructor, it may be tempting to use the simulator in the same way as you teach in the airplane; however, this does not maximize its effectiveness. To get the most out of the time you spend in the simulator, you should follow some straightforward principles.

APPROPRIATE USE

The most important single principle of using a simulator in flight instruction is to use it appropriately. That is, use it for what it is good for. Do not try to make it perform in a way for which it is not suited. For example, if you want to teach a beginning student about the fundamental relationships between control inputs and how they affect the airplane, you can use almost any simulator, high- or low-fidelity. The reason is that the student does not know enough to be able to distinguish between high-fidelity and low-fidelity simulators. You are not trying to teach the student how to fly very precisely at this stage, but rather are forming a mental picture of the relationships. You can have the student fly straight and level, perform turns, climbs, and descents, or see the effects of power on airspeed in the simplest of simulators.

On the other hand, it is most probably inappropriate to use a low-fidelity simulator for teaching an experienced pilot certain instrument procedures, such as an ILS approach. The pilot will be distracted by the performance of the simulator and may not be able to concentrate on what you are teaching.

PROFESSIONAL USE

The second principle for making simulators effective is to use them professionally. There are two parts to this. The first is to have a positive approach about their use. The second is to be as prepared for a simulator lesson as you would be a flight lesson.

Quite often, students react negatively toward ground training devices. This may be a reflection of the way they perceive the value of simulation time versus airplane time. You may have to convince the student that their time is well spent. Unfortunately, this negative attitude may come from other flight instructors who would rather be flying an actual airplane. It is important to remember that student attitudes toward simulation will be influenced by your own. The more positive you are, the better the students will regard their time.

The second aspect of using simulators effectively is to be prepared. As we mentioned in the previous section, it is important to prepare your flight lessons carefully beforehand. This enables the student to obtain maximum benefit from the time spent in training. The lesson plan for a simulator should include notations about the simulator settings for wind, position, radios, and so on, as well as all the items on a normal lesson plan. [Figure 2-3] An additional benefit from being well prepared is that you convey the message to your students that you regard simulator time as valuable.

EFFECTIVE USE

The third principle for making the most of simulator training is to use them effectively. What this means is to use the features and facilities of the simulator in a way that enhances the student's learning. Typically, these features include the ability to position the simulator anywhere in space at the touch of a button, to freeze the simulator at any time, to control environmental factors (wind, clouds, and turbulence), and sometimes to replay a portion of the flight.

If you use these features properly, you can increase the productivity of a training session. For example, you can often practice many more approaches in an hour in a simulator than you can in an airplane. This is accomplished by repositioning the airplane near the approach rather than having the pilot fly it there.

LESSON __VOR/ILS Instrument Approaches__ STUDENT ___ DATE ___

OBJECTIVE
- To develop the student's instrument flight proficiency in VOR, localizer-only, and ILS approaches.

ELEMENTS
- Review
- Straight-in VOR Approaches
- Circling VOR Approaches
- Localizer Approaches
- ILS Approaches
- Partial Panel

PRESETS
- Wind   120/05
- Turbulence   1-2
- Engine Sound   4

SCHEDULE
- Preflight briefing   : 10
- Review constant airspeed descents/ constant rate descents   : 10
- VOR RWY 9 — Auburn, IN (From IAF)   : 10
- VOR-A — Beach City, OH (From IAF) Repeat   : 20
- ILS RWY 27 — Lima, OH (Localizer-From FAF) Repeat (Glide Slope)   : 15
- Vacuum pump failure
- ILS RWY 22 — Evansville, IN (Localizer Only) Repeat (Glide Slope)   : 15
- Postflight critique   : 15

EQUIPMENT
- Brand B-220 Flitematic

INSTRUCTOR'S ACTIONS
- Discuss lesson objectives
- Configure trainer
- Position trainer
- Present malfunction
- Critique student performance

STUDENT'S ACTIONS
- Discuss lesson objectives
- Perform maneuvers and approaches
- Respond to malfunction
- Ask questions

COMPLETION STANDARDS
- The student will learn the procedures necessary for performance of straight-in and circling VOR approaches, localizer-only approaches, and ILS approaches. The student will demonstrate the use of power and attitude changes to control airspeed and descent rates.

Figure 2-3

Properly used, simulators can also provide more thorough training than is possible in an airplane. Because you have control over the environment, you can ensure that a student has proper crosswind landing experience, for instance. If you had to rely on the weather for this, you may never be able to find all possible conditions. Similarly, you can create equipment malfunctions safely in a simulator, such as an engine failure on takeoff, that might create a dangerous situation if you were to try the same things in an airplane.

CONCLUSION

Learning to fly is a wonderful experience that can be enhanced by appropriate study skills and learning environment. With the proper approach, your learning can be more effective, lead to longer retention of the material, and be less expensive, which provides more opportunities to fly.

In this chapter, we have outlined ways you can improve your learning of both ground school and flight materials. We have also looked a little into the future to give you an idea of topics you may encounter in the near future, as well as techniques of teaching that can benefit you.

Exercises

Chapter Questions

1. Explain why learning to fly is considered to be a good example of where human factors can be applied.

2. What is the best way to learn?

3. List five techniques that can improve your ability to learn effectively. Briefly comment on why each technique works.

4. Explain why flying is primarily a mental pursuit, not a physical one.

5. Make a set of notes about this chapter.

6. Draw a concept map on one of the following topics: choosing a place to learn to fly, techniques for improving learning, and human factors and learning.

7. Prepare five questions on topics in the preceding question that confuse you. Give them to an instructor and ask for help in understanding them.

8. Describe why structured practice can help you to learn more effectively.

9. Prepare a list of questions to determine how well a ground school/flight school covers the topic of human factors.

10. Why are simulators effective in flight training?

11. Explain how computers can be used to improve your learning.

12. Give five principles of good ground school instruction. Do the same with good principles of flight instruction.

REFERENCES AND RECOMMENDED READINGS

Alessi, S.M. & S.R. Trollip. 1991. *Computer-based Instruction: Methods and Development (2nd ed.).* Englewood Cliffs, NJ: Prentice Hall.

Armbruster, B.B. & T.H. Anderson. 1984. "Mapping: Representing informative text diagrammatically." In C.D. Holley and D.I.Dansereau (Eds.), *Spatial learning strategies: Techniques, applications, and related issues.* New York: Academic Press.

Bruner, J.S. 1966. *Towards a Theory of Instruction.* Cambridge, MA: Harvard University Press.

Federal Aviation Administration. 1977. *Aviation Instructor's Handbook.* FAA publication AC 60-14. Washington, D.C.: U.S. Government Printing Office.

Gagne, E.D. 1977. *The conditions of learning (3rd ed.).* New York: Holt.

Mager, R.F. 1962. *Preparing Instructional Objectives.* Belmont, CA: Fearon Publishers.

Private Pilot Manual. 1990. Englewood, CO: Jeppesen Sanderson, Inc.

Cockpit Design **3**

INTRODUCTION

You may wonder why we have a chapter on cockpit design in a book on human factors for general aviation. An easy answer is that the cockpit is your work place when flying, and human factors is the study of the interaction of humans and where they work. Specifically, there are three reasons why you should know something about the design of cockpits. First, the cockpit is your major source of information as a pilot and the means by which you control the aircraft. Consequently, its design has a major impact on your performance. Second, the nature of the cockpit is changing rapidly, and it is helpful to understand these changes in light of human factors' norms. Third, sometime in the future, you may want to buy new equipment for your airplane. Having a basic idea of design will help you compare competing products in terms of their human factors design.

In this chapter, we introduce the cockpit as an integral part of the system of flying and discuss its major components: namely, the displays and controls. We emphasize how the design of each of these affects your ability to perform well. Later in the book, we introduce some of the psychological issues of design and discuss how they relate to airplanes.

THE MAN-MACHINE SYSTEM

We start our discussion by introducing you to the simple concept of a basic man-machine system, which is what a pilot and an airplane represent. The definition of a **man-machine system** is a collection of components, some of which are machines and at least one of which is a human, that work together to accomplish some purpose.

We have had man-machine systems since the first cave man picked up a rock to throw at enemies or made a spear to assist in the hunt for food. These are examples of **open-loop man-machine systems**. The system is called an open-loop because there is no control provided to the user. For example, once the spear has left the cave man's hand, he can do nothing about it. The only feedback is watching the spear as it flies toward the animal.

Later in history, one of the first **closed-loop systems** was developed when animals were domesticated and harnessed. A closed-loop system adds continuous control and feedback to the human operator. For example, if you are riding a horse and it strays away from the desired path, you are able to adjust its direction by applying pressure to the reins. The reins provide the control. The position of the horse with respect to the desired path is the feedback.

Another example of a closed-loop system is the thermostat and furnace which maintains your house at a given temperature in winter. When you set a desired temperature on the thermostat, the heating system engages if the house is too cold. This causes hot air to circulate, and soon the house warms up. The thermostat monitors the temperature. When the desired temperature is reached, the thermostat turns off the furnace. The temperature now starts to drop, and the cycle repeats itself. The system is closed-loop.

The reason for introducing these concepts is to draw your attention to the basic design differences in each. In an open-loop system, it is not necessary to spend much effort in designing the information that is sent back to the human, because he or she has no control once the system is set in motion. However, in a closed-loop system, the human needs accurate information to know what is happening with the system and good controls are required to manipulate the system accurately.

Figure 3-1 is a block diagram of a closed-loop man-machine system such as an automobile. You can see that the basic elements include an input (such as a need to drive some-where), displays (or instruments), the man (or human controller), the controls (steering wheel, gas peddle, brakes, etc.), the machine (automobile), feedback (information from the environment and automobile such as speed, direction, position, etc.), and the output (transportation to desired location). From this diagram, you can see why it is called a "closed-loop" — the feedback is a return "loop" which makes it a continuously controlled system.

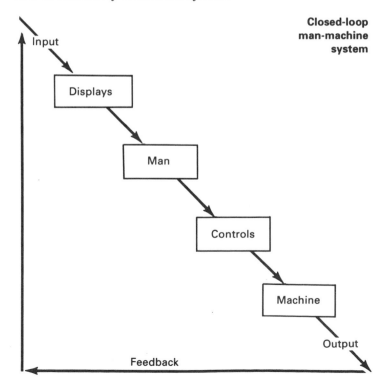

Figure 3-1

Consider the basic man-machine system in the airplane. [Figure 3-2] Notice that information flows from the flight controls (rudder pedals and control wheel) to the airplane (rudder, ailerons, and elevator) and from the airplane to the displays, where it is presented to you, the pilot. Information also flows between you and the displays, because you have to set the displays properly to get the correct information. Information also flows both ways between you and the controls. You move the controls to change the attitude of the airplane, and they provide feedback concerning the status of the airplane through the amount of pressure you feel.

The airplane is a closed-loop system. You provide input to the airplane; the pressure on the controls and the readings on the gauges provide feedback to you. Based on this information, you modify the input, continuously trying to attain the desired results. Because it is a closed-loop system, it is very important that the system provides accurate information and that you have a sensitive and accurate means of control.

The two major parts of the airplane with respect to the man-machine system are the gauges, or displays, and the controls. Later in the chapter, we discuss both of these and their relationship to you, the pilot. We pay particular attention to how their design influences how well you fly.

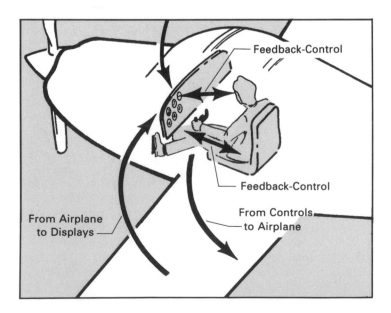

Figure 3-2

DESIGNING MACHINES FOR PEOPLE

Traditionally, engineers have designed machines according to engineering principles, rather than behavioral principles. That is, when a machine has been designed, very little consideration has been given to how easy it is for a person to use or operate. Life is filled with such examples. In some cases, the poor design is a nuisance and not particularly dangerous, such as restroom mirrors that are not high enough or spare tires in automobiles that require you to unpack the entire trunk to access. In others, the design is positively dangerous, such as automatic transmissions in some cars that unexpectedly engage or, in our field, the design of cockpits that cause pilots to make errors.

We believe that there is a simple reason why engineering principles have dominated the design field. It is that many engineers who have not had the benefit of human factors training believe that humans are able to adapt readily to almost any environment in which they are placed. That is, at least on the surface, humans can operate almost any machine no matter how it is designed. Whether a work place is too hot or too cold, too quiet or too noisy, humans manage to perform their jobs. The real question, however, is whether they perform these jobs adequately and safely, and whether another design would have resulted in better, more economical and safer performance. We are convinced that enormous productivity losses have occurred through the bad design of machines, and even worse, many lives have been needlessly lost.

In the aviation world, the result of engineering-dominated design has been airplanes that are difficult to fly, that require constant relearning, and that cause pilots to make errors. All of these effects are costly both in terms of money and human suffering.

For example, for years a well-known manufacturer of general aviation aircraft was out of step with the rest of the industry and designed its airplanes with the throttle and propeller controls, as well as the gear and flap handles, in the reverse position to everyone else. Because pilots normally fly a variety of airplanes, they have trouble with the older airplanes of this manufacturer, particularly with respect to the gear. In fact, this is so much so, that the rate of inadvertent gear-up landings in one of the models is much higher than it is in other retractables. When the manufacturer finally made the switch to the industry standard, pilots who were familiar with their older models had problems because of the mental set that they had established for the original configuration.

One pilot, for instance, who for years owned a twin made by the manufacturer and seldom flew other airplanes, bought a later model of the twin with the new positions of gear, flap, throttle, and propeller controls. He received a checkout in the new airplane and had no trouble with the new position of the controls until he was landing in the Bahamas with a load of fun-seeking friends. This time, as he was rolling out after landing, he reached to raise the flaps and moved the gear handle to the retracted position. You know the rest of the story.

In accidents like this one, investigation reports often cite "pilot error" as the cause. However, in our opinion, this term can be misleading, because in many such cases the pilot was led into the error by the design of the equipment. In our estimation, the error was not caused by any lack of vigilance or preparation on behalf of the pilot, but rather by a design that violated human factors principles.

In contrast to engineer-driven designs, the goal of human factors is to design systems to minimize the need for training and to minimize the probability of mistakes. Human factors-driven design also concerns itself with developing and providing appropriate training to complete any job left undone in system design. In aviation, we believe that if airplanes were properly designed, there would be fewer accidents and training could be greatly reduced.

However, we cannot readily reconfigure airplanes that already exist; all we can do is draw your attention to the issues of design that may adversely affect your performance. Our hope is that this increased awareness of potential problems will lead you to make fewer mistakes while flying.

DISPLAYS

After World War II, an eminent aviation psychologist, Dr. Paul Fitts, was very concerned about the number of errors pilots were making while flying military aircraft. To gather further information, he conducted a survey of pilots, asking them to list the errors they had made or had observed other pilots make while flying. Fitts tabulated the results into a classification scheme for the purpose of improving the design of both controls and displays in cockpits.

Using these error data and taking into account the scan requirement in the cockpit, Fitts developed the "Basic-T" for the location of the flight instruments, a configuration that is still used today in virtually all aircraft. [Figure 3-3] The central instrument is the attitude indicator, a gyroscopic flight instrument. It is a substitute for the horizon and is, therefore, the primary instrument used when flying in poor visibility conditions. The purpose of the attitude indicator is to provide you with an immediate and direct indication of the airplane's pitch and bank. In fact, it is the only instrument that can do both.

Surrounding the attitude indicator are the instruments that provide other information about the airplane. The airspeed indicator, to the left of the attitude indicator, provides information about the speed at which the airplane is moving through the air.

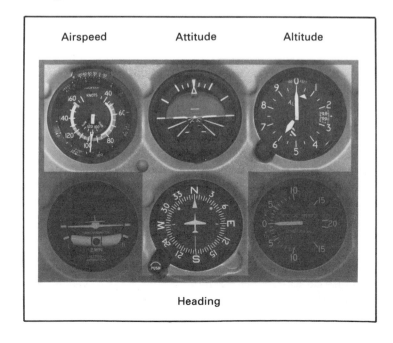

Figure 3-3

To the right of the attitude indicator is the altimeter, a pitot-static instrument. If set properly, it gives a direct reading of the airplane's altitude relative to sea level.

Finally, the heading indicator is placed below the attitude indicator. Like the attitude indicator, it is a gyroscopic instrument. It provides direct information as to the heading of the airplane.

Fitts believed (quite correctly) that the attitude indicator is the most important instrument for flying safely in conditions of poor visibility. Consequently, it is placed at the center of your vision. The primary, detailed information you need to monitor your flight (altitude, airspeed, and heading) is clustered around the attitude indicator.

Fitts' "Basic-T" was one of the first attempts to apply human factors principles to cockpit design, and it has certainly benefitted pilots. However, the instruments themselves are often examples of poor design.

THE ATTITUDE INDICATOR

The attitude indicator — the most important single instrument — is a good example of poor design. It is generally accepted that a good design principle is one where the moving element of the display moves in the same direction as the control to which it is related. This is known as **control/display compatibility**. For example, if you put the flaps down, the indicator should go down. This principle seems so logical that it is hard to imagine that it could be violated in an airplane. Surprisingly, it is violated in three prominent cockpit displays, the attitude indicator, the course deviation indicator, and the magnetic compass. Because these instruments lack control/display compatability, they have a "direction of motion" problem.

The moving element of the attitude indicator is the moving horizon. When you move the control wheel to the right, the horizon moves to the left. When you pull the wheel back to climb, the horizon moves downward. Both indications are opposite the direction of motion of the control. The reason for designing the attitude indicator this way is that the horizon element accurately mimics the real horizon (and it

Inside-out   Outside-in   **Figure 3-4**

is easier to do with a mechanical instrument). If you move the wheel to the right, the airplane rolls right which gives the impression to the pilot, who is now at an angle to the ground, that the horizon has rolled to the left.

However, this is an engineering reason and not based on normal human reactions. A quick confirmation of this direction of motion problem can be seen in the difficulty most student pilots have when the attitude indicator is explained. The instructor will usually tell you that the indicator's horizon moves in the wrong direction unless you think of it in terms of a view out of a hole in the front of the aircraft. Viewed as such, you will see the horizon move rather than the airplane when you move the control. This is why the display is called an **inside-out** display — it reflects what you would see out of the cockpit window. [Figure 3-4, left side]

Compare this inside-out attitude indicator with one in which the airplane symbol moves in the same direction as the control movement, called an **outside-in** display. [Figure 3-5, right side] The two indicators provide the same information in different ways. Research has shown that nonpilots are able to control an airplane better with an outside-in presentation than with the inside-out one used in the conventional attitude indicator. Of course, after experience with the inside-out display, pilots learn to overcome their natural tendencies to interpret it incorrectly. However, there are still accidents that are caused by pilots misinterpreting the attitude indicator and rolling the wrong way — usually upside-down.

Figure 3-5

THE ALTIMETER

The altimeter is another instrument whose design causes problems. The three-pointer altimeter has been the design of choice for many years. [Figure 3-5] However, it has a major drawback. There can be confusion between the small and medium size needle, leading to a reading 10,000 feet in error. Such reading errors have resulted in a number of accidents and incidents.

Attempts have been made to solve this problem in large aircraft by using a vertical tape, whose vertical position represents altitude. The disadvantage of the vertical display is that the tape is so long that only a part can be shown. [Figure 3-6] This means that the pilot must read the number on the tape to determine altitude rather than being able to determine it from the length of tape.

Tape Altimeter

Figure 3-6

Analog   Digital   **Figure 3-7**

This raises another interesting issue: namely, the difference
between digital and analog displays. The best example to
illustrate the difference is a wristwatch. Watches with sweep
hands are **analog** devices, because the position of the hands
provides the time on a continuous scale. [Figure 3-7] **Digital**
watches give the time in numeric form. Although digital
watches are more accurate, they actually provides less
information than their analog counterparts. It is much easier
to sense amounts of time, such as the time left to the end of
a class period, on an analog watch than it is on a digital
watch, which requires a mental computation.

Similarly, even though a digital readout of altitude provides
you with more accurate information, it provides no visual
representation of how far you are off a particular altitude. To
obtain this information, you have to make a mental calculation. For example, if your target altitude is 11,500 feet, it is
easier to tell from figure 3-5 that you are a little high than it
is from figure 3-8, which represents a true digital altimeter.

Figure 3-8

HEADING INSTRUMENTS

A comparison of the heading indicator and magnetic compass provides an excellent study as to why design issues can lead pilots into making errors. The heading indicator in most modern airplanes conforms to the control/display compatibility principle mentioned above. As shown on the left side of figure 3-9, if you want to turn to a heading of North, you must turn toward North on the display: namely, right. The display leads you to the correct instinctive decision on how to act.

On the other hand, the magnetic compass on the right side of figure 3-9 works in exactly the opposite manner. In our example, the magnetic compass display shows North to the left of your current heading. Now imagine being in a stressful situation in clouds with an inoperative heading indicator. You must now rely on the magnetic compass. Radar control is vectoring you for an approach to the nearest runway. Every time you are given a new heading, you have to think about which way to turn, because you are not used to the reversals inherent in the compass. The more promptly you have to respond to ATC's directions, the more dangerous these delays become.

Heading Indicator

Magnetic Compass

Figure 3-9

VOR   HSI   **Figure 3-10**

VOR AND HSI

Another good contrast lies in the traditional VOR display and
the Horizontal Situation Indicator (HSI). [Figure 3-10] The
traditional VOR display violates the control/display compatibility principle. For example, if your airplane moves to the
right of course, the course deviation indicator moves to the
left. Furthermore, there is no relationship between the
aircraft heading and the reading on the VOR; that is, any
heading yields an identical display.

On the other hand, the HSI has improved on the VOR in two
ways. First, the compass card is slaved to a remote gyro-
scope, so the heading of the airplane is continuously reflected in the display. Second, although the course deviation
indicator still violates the control/display compatability
principle, the more pictorial display helps you interpret
where you are relative to where you want to be. The small
airplane on the display points in the direction you are
actually going.

In the cockpits of many of today's commercial and larger
general aviation aircraft, the instrument panels contain CRT's
to display the information. [Figure 3-11] These are usually
called **glass cockpits**. Being computer-controlled, their
displays can be designed more easily with human factors
principles in mind. Figure 3-12 shows the electronic equivalent to the HSI. It improves the display once again, so that
misinterpretations are rare. We discuss glass cockpits in
depth later in the book.

Figure 3-11   *Photo courtesy of Collins Avionics, Rockwell International Corp.*

*Photo courtesy of
Collins Avionics,*
Figure 3-12   *Rockwell International Corp.*

PROGRESS IS SLOW

The unfortunate thing about design is that we seem not to learn from our mistakes — that we have not built on what Fitts started. For example, the Flight Management System (FMS) used in modern, high-performance general aviation aircraft as well as airliners, appears to have been designed on engineering principles rather than taking human factors into account. In particular, the FMS has a keyboard that contradicts what we know about how humans function. First, it is unlike any other keyboard that pilots may use and, second, the keys seem to have been positioned considering economy of space, rather than ease of use. As a result, it is very difficult to learn and is highly error prone.

CONTROLS

In the previous section, we discussed displays, which are your major source of information regarding the progress of the flight. We now look at aircraft controls, which are the means by which you can change the current configuration of the airplane. Controls are the action part of the aircraft/pilot interface. Most people are quite good at controlling vehicles using various parts of the body, such as the hands, feet, voice, and fingers.

TYPES OF CONTROLS

Because of the varied nature of the piloting task, almost every type of control known has been tried in the design of aircraft cockpits. They cover the full gamut from controls requiring large hand movements to speech controls. In the future, they may even include controls activated by thoughts. One reason for varied types of controls is that the piloting task requires several coordinated inputs applied simultaneously. Proper design enables this to happen. For example, on many airplanes, a single hand can simultaneously bank the airplane and change its pitch, as well as adjust the trim and activate the microphone. At the same time, you can be using the other hand to change the power setting, your feet for rudder input, and your voice for communications.

There are both large and small controls in airplanes. The large ones include the control wheel or stick, the rudder pedals, the flap and gear controls, throttle, propeller, mixture, carburetor heat, and cowl flaps. Each of these require the use of hands or feet and, in some cases, fairly large amounts of force. The movements are either forward and back, left and right, or twisting motions.

The small controls include the radio, the push-to-talk, panel light, and autopilot switches, as well as RNAV, LORAN C, and the flight management computer keyboards. Other examples are the knobs for setting the altimeter and artificial horizon, the audio panel switches, the landing light, pitot heat and other electrical switches, the cockpit and cabin environmental control switches, and the circuit breakers. These controls require finger movement with either push-pull or twisting movements.

CONTROL DESIGN

Pilots often have to find and manipulate controls without looking at them. In fact, you are encouraged to do so for night flying. Consequently, it is necessary to be able to distinguish one control from another by its shape and position, although color is sometimes also used.

Human factors specialists have shown that if controls have a form or shape that roughly represents the aircraft system being controlled, you will be more likely to find and move the right control. If, in addition to this shape coding, the controls are located in the same place on all aircraft, you will form a mental set that minimizes the chance of error.

For example, the gear handle has a round shape with an indentation on the side like a wheel. The throttle and mixture controls are color-coded black and red respectively and are of different sizes, because it would be dangerous to pull back the mixture instead of the throttle. The propeller control is shaped with a type of scalloped edge, which represents a propeller. And electric flap handles are usually shaped like an airfoil. [Figure 3-13] Each of these shapes and colors are determined deliberately using known human expectations so as to enhance the development of safe habits.

In the example of the inadvertent gear retraction discussed previously in this chapter, the manufacturer wanted to have the same handle movements for both the flaps and the gear. To accomplish this, they needed to reverse the recommended position of the airfoil. They designed it with the leading edge outward instead of the recommended and industry-standard orientation of the trailing edge out. This made the flap handle indistinguishable from the gear handle.

Figure 3-13

There is a lesson to be learned from this. When you are flying an unfamiliar airplane with retractable gear, you must exercise caution when using the flap and gear handles. Give yourself plenty of time to correct any error you may make.

SOUNDS AND ALARMS

There are numerous alerting systems on board both general aviation and airline aircraft today. These are sometimes referred to as "bells and whistles." These can be aural, visual, or both. Their purpose is to alert you of an impending dangerous situation. For example, all airplanes have stall warning devices to alert you that the angle of attack is approaching a stall. And, if you are flying an ILS, there is a blinking light and a beeping sound to indicate that you are over the marker beacons.

There are three problem areas in general aviation with respect to aural alarms. The first is familiarity. That is, if an alarm is heard frequently, it can easily be ignored. For example, some retractable gear airplanes have audible warnings to alert the pilot that the gear is up when the throttle is reduced below a particular setting. If you are practicing power-off stalls, this alarm will go off every time you reduce power. Soon, you will start to ignore it and may well not hear it if it goes off in a real gear-up landing situation.

The second problem area occurs when alarms are temporarily disabled because they are too intrusive. In the case above, when practicing a maneuver that activates an audible alarm, it has been known for the instructor or pilot to facilitate communication by disabling the alarm by pulling a circuit breaker. It sometimes happens that the instructor or pilot forgets to reset it.

A third problem area with respect to audible alarms is that information coming through your ears can be overwhelmed. That is, if you are concentrating hard on some task, perhaps an emergency, it is possible for an alarm to go off unheard. That is, the depth of concentration prevents you from hearing anything.

The best way to deal with these problems is to form good habits. Train yourself to actively check the cause of every audible alarm, even if it has been going off frequently. In this way, you will not be caught napping by a real emergency. Second, if you ever disable an alarm, try to leave some indicator, such as a red ribbon tied to the control wheel, to remind you to reset it.

CONCLUSION

You and your airplane make up an aviation system. In such a system, it is important that you obtain good information quickly and accurately and control the airplane in the same way. In this chapter, we have illustrated a number of cockpit design issues that can affect how you perform. The more you understand how design affects your performance, the better prepared you will be to deal with the variety of designs in general aviation. Later in the book, we deal in depth with the principles of design, particularly in the context of the increasing automation of aircraft.

Exercises ▪▪▪▪▪▪▪▪▪▪▪▪▪▪

Chapter Questions

1. Why is cockpit design so important in preventing pilot errors?

2. What can you, as a pilot, do to prevent errors in a cockpit that is not well designed from the human factors perspective?

3. Why is it important that different airplane manufacturers standardize their placement of controls in an aircraft cockpit?

4. What are the essential elements of an open-loop, man-machine system?

5. What are the essential elements of a closed-loop, man-machine system?

6. What is the value of color and shape coding of controls? Give a good and bad example of each.

7. What is meant by the "control/display compatibility" principle? Give some examples of its violation in the cockpit and the potential consequences.

8. Give five examples of poor design in your home or office. How would you remedy them?

REFERENCES AND RECOMMENDED READINGS

Diehl, A.E. 1981. General Aviation Cockpit Design Features Related to Inadvertent Landing Gear Retraction Accidents. In R.S. Jensen (Ed.) *Proceedings of the First Symposium on Aviation Psychology*. Columbus, OH: The Ohio State University Department of Aviation.

Jensen, R.S. (Ed.) 1989. *Aviation Psychology*. Brookfield, VT: Gower Technical.

National Transportation Safety Board. 1967. Aircraft Design Induced Pilot Error. Special Study PB 175629.

The Eyes and Ears

4

INTRODUCTION

Early in the book, we described a simple information processing model that helps us to understand how a pilot functions in the cockpit. The first part of that model deals with the information a pilot needs and how it is gathered. The second part concerns how the information is processed by the brain. The third part relates to the decision-making process, and the final part deals with the implementation of the decisions. In this chapter, we look at the first part of the model and discuss the brain's two most important means of obtaining information: namely, the eyes and ears. It is through these sensors that by far the greatest proportion of information is made available to a pilot.

It is important to note that information gathered by both the eyes and the ears can be distorted by the brain, leading you to perceive erroneous or misleading information. Being aware of these possible distortions will enable you to take preventative or precautionary action to minimize their effects.

THE EYES

The discussion of the eyes would be simple if they operated like a camera, where light passes through a lens and is captured on the "film" at the back (the retina). If this were the case, the role of eyes would be easily understood.

In reality, vision is an extremely complicated process that involves much more than the eyes themselves. It also involves the brain, which processes all visual inputs. What this means is that the scene or object you are looking at may be quite different from what you think you see. The original scene can be distorted both by the physical attributes of the eye and the way the visual information is processed.

For example, look closely at figure 4-1. Do you see shadows in the white areas at the corners of the black boxes? This is

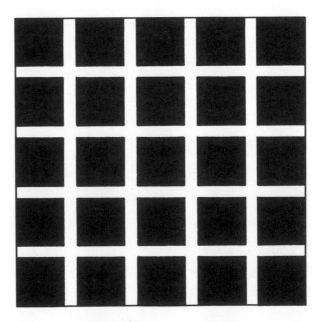

Figure 4-1

Figure 4-2

a normal reaction. In reality, there are no shadows, although the light sensors in the eye create this illusion. In this case, the physical attributes of the eye cause a distortion.

On the other hand, the brain itself can distort reality. To see what we mean by this, look at figure 4-2 and read the message in the triangle. What most people read is "Bird in the hand." They do not see that the word "the" is repeated. A simple explanation for this is that the phrase "bird in the hand" is so familiar that as soon as the brain processes "bird in the" on the top two lines, it expects to see "hand" on the third. When, in fact, "hand" appears on the third line, the brain ignores or forgets the extra "the." The final message understood by the brain is the well known phrase with only one "the."

As pilots, it is important to know how visual information can be distorted, both physically and mentally. It is also helpful to know the situations in which distortions are likely to occur. An awareness of these factors may help you consciously compensate for them and, hence, fly safer.

PHYSICAL ISSUES

Even though the details of the visual system are extremely complex and not completely understood, it is useful to have a basic understanding of how the eye is constructed and how it works. Figure 4-3 depicts the parts of the eye. As you read through this section, you may find it helpful to refer back to this diagram.

As you read this page, the visual system is allowing the light reflected off the page to fall on your retina. First, the light passes through the lens near the front of your eye. The amount of light is regulated by the size of the pupil. When there is little light, the pupil enlarges to allow more light to strike the retina. When the light is bright, your pupil contracts to protect the retina. In this way, it is very much like the aperture on a camera.

ACCOMMODATION

The second part of the process is the focusing of light on the retina. To do this, the eye works differently from a camera, even though the effect is much the same. The lens in a normal eye is convex, which means that light reflected from an object is bent as it passes through the lens. The point at which the rays of light intersect, or are concentrated, is the focal point. When this point coincides with the retina, the object is in clear view, or more simply, it is in focus. An object is out of focus whenever the focal point is in front or behind the retina. In the latter case, the rays of light actually strike the retina before they reach their focal point.

The eye, however, can change its focus by changing the shape of the lens. If the focal point lies behind the retina; that is, it needs to be moved forward so it falls on the retina,

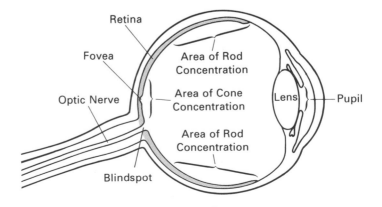

Figure 4-3

the lens is made fatter by a contraction of the muscle surrounding it (the ciliary muscle). If the focal point falls in
front of the retina; that is, it needs to be moved farther back
so it falls on the retina, the lens flattens. This process of
changing the shape of the lens is called **accommodation**. In
a camera, on the other hand, focusing is accomplished by
moving the elements of the lens closer to or farther from the
film.

CORRECTION

When people grow older, one of the natural effects is a
stiffening of the lens. This makes it more difficult for the
ciliary muscles to change the shape of the lens. To compensate for this, we wear glasses or contact lenses to bend the
light before it hits the lens. The lens then is able to make the
necessary fine adjustments so the image is in focus on the
retina. The left-hand diagram in figure 4-4 depicts an eye that
is shortsighted. That is, light from a distant object is focused
in front of the retina causing the image to blur. If the eye's
lens cannot compensate for this, a concave lens is placed in
front of the eye, bending the light outward. The light then
passes through the lens and focuses on the retina. This is
shown in the right-hand diagram of figure 4-4.

RODS AND CONES

When light falls on the retina, it activates two types of light-
sensitive cells called rods and cones. It is an important fact
to understand that the rods and cones are not evenly distrib-
uted in the retina. A small area in the center of the retina
(called the fovea) is covered entirely by cones. From the center to the edge of the retina, the number of cones decreases
and the density of rods increases. At the edge, there are only
rods.

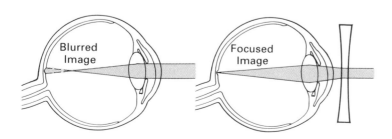

Figure 4-4

Rods and cones have very different characteristics. Cones are color sensitive and are used for fine discrimination. They require good lighting conditions to function properly. Cones are activated when you look directly at something. On the other hand, rods are most useful in low-lighting conditions. They are not color sensitive and provide most information about objects in the periphery of vision.

Low-Light Conditions

What does this mean for the pilot? The greatest impact comes when the light is diminished, such as at dawn or dusk, or at night. Because the cones at the center of the retina work best in bright light, their functioning is adversely affected by low-lighting conditions. Both acuity and color perception require the use of the cones. When light is low, for example, it becomes increasingly difficult to read printing that is easy to read in bright light, even when there is enough light to see every letter. In some cases, people who would not normally need corrective lenses for daylight, may find it difficult to decipher writing in low-light conditions.

Another aspect of vision that is adversely affected by low light is the ability to see other airplanes. As we mentioned above, the fovea is covered completely by cones, which creates a blind spot in low-light conditions. It is not very effective, therefore, to search for a target at night by staring directly at it. It is better to use your peripheral vision. If you think you see the position or anti-collision lights of an airplane against the lights of a town, for example, do not stare at the point where you thought you saw the airplane. Rather, look 10 or 15 degrees away from it and use your peripheral vision to pick it up more readily.

The other effect of low-light situations is in the ability to perceive color. The cones at the center of the eye provide most of our color information. However, they require bright light to function most effectively. This is the reason why everything loses its color at dusk. A garden full of brilliantly colored flowers has a very uniform drab color as the sun goes down. For the pilot, anytime color is used to transmit information (such as terrain on a sectional chart or lines on an airspeed indicator), great care must be exercised. In low-light situations, important color differences may not be discernable. The situation often becomes worse at night because the cockpit usually is illuminated with red light which can mask important color distinctions.

Color Blindness

For pilots who suffer from color deficiency (sometimes called color blindness), low-light situations can make color discrimination even more difficult. For example, it may be very hard to tell the difference between the red and green position lights on an airplane's wings (which means you do not know in which direction the airplane is heading) or between differently colored lights on an airport (making it easy to confuse runways with taxiways). Color blindness affects a person more when flying than when driving. There is not much difficulty in driving because there are few color cues that are required for safe performance. The major source of color-coded information is, of course, the traffic light. The position of the three colors is constant; therefore, position can be used as a practical substitute for color. This is not the case in flying. The position of taxiways and runways are different from one airport to another, so you cannot predict where they are from the position of the lights. You have to rely on their color.

It is worth mentioning at this stage that the functioning of the eyes is dependent on the amount of oxygen available. If you are starved of oxygen, vision deteriorates. This is an important fact because it means that both smoking and hypoxia are likely to have an adverse effect on your ability to see properly. If you like to fly at 5,000 feet AGL or above and are a heavy smoker, there is a good chance that your vision may be severely impaired. Unfortunately, one of the other effects of hypoxia is a feeling of well being, so you may not be inclined to take corrective action even if you notice the symptoms.

EMPTY-FIELD MYOPIA

We mentioned earlier that the ciliary muscles in the eye allow us to focus on objects at different distances. In situations where objects are not clearly defined, such as in haze, the eye has a tendency to revert to a resting position called "empty-field (or empty-sky) myopia" without you realizing it. Although this distance varies considerably between individuals, it is usually at about 3 to 5 feet (1 to 1.75 meters). The effect of empty-field myopia is to make objects appear smaller and, hence, farther away.

To illustrate this, stand about 15 feet (5 meters) from an object that is about 3 feet wide (1 meter), such as a door. Look at it and sense its width. Now, extend your arm and raise a finger so that the finger is about 20 degrees to one side of the door. Focus on the finger. In the periphery of your vision you will still see the door; however, it will appear to be a little narrower than when you were looking directly at it. When flying, objects such as distant mountains can appear farther away if your eyes have reverted to their empty-field state.

Another effect of this reversion to empty-field state is for objects to be out of focus. Whenever this happens, you lose the ability to see detail. This can be dangerous because it is hard to pick out other airplanes at a distance since they occupy such a small area of your visual field. Sometimes, you do not even see them.

You can illustrate this effect quite easily. Hold this book at arms length and focus on the text. You should be able to read it clearly. Now, shift your vision slightly to one side of the book and focus on something behind it — even a few inches is sufficient. Notice that the writing on the page is now impossible to read. In fact, you cannot even see the spaces between words. Similarly, if your eyes are at empty-field state while flying, it is very difficult to notice traffic, because it melds into the background.

If you are flying in a situation conducive to empty-field myopia, it is good practice to focus quite frequently on your airplane's wingtips. These are far enough away to break the empty-field state. Objects that are closer, such as instruments in the cockpit, are usually so close that they do not affect the empty-field myopia.

BLINDSPOT

The final purely physical aspect of the eye that we want to discuss is the **blindspot**. The optic nerve carries all the visual information from the eye to the brain. Where it connects to the eye there are no rods or cones, which means you are completely blind in this spot — hence the name. To illustrate this, close your left eye and look at figure 4-5. Hold the book at arm's length and focus on the low-wing airplane. Slowly bring the book toward you. At some point the high-wing airplane on the right will completely disappear. You can repeat this with your left eye by focusing on the high-wing airplane. In this case, the low-wing airplane will disappear.

Under normal circumstances, your binocular vision eliminates the blindspot, because an object cannot be in the blindspot of both eyes simultaneously. However, if your field of vision is partially obscured, it is possible for one eye to be unsighted, which may lead to an airplane being in the other eye's blindspot. The obvious implication is that it is essential for you to keep scanning both the instruments and the environment. If you fixate on any single point (for example, a mountain in the distance), it is possible you will not see an airplane on a collision course with you because it is in your blindspot.

There is actually another type of blind spot that is more a function of airplane design than the eye itself. This occurs when traffic is obscured by a door post, wing, or fuselage. Merely turning the head may not reveal traffic obscured in this way. You should also move your head in such a way to see behind all obstructions.

In this section, we have discussed how some of the purely physical attributes of the eye affect a pilot's performance. Later in the book, we discuss how changes in the physical condition of the pilot's body affect the ability to see. In particular, we will look at the effects of fatigue, hypoxia, and smoking on vision. Before we reach that section, however, we want to discuss the role of the brain in vision and some of the distortions the brain can cause.

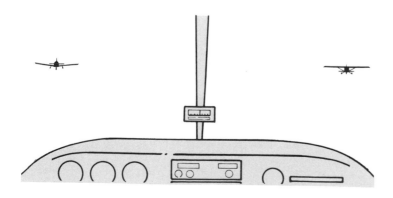

Figure 4-5

PSYCHOLOGICAL ISSUES

What you see is a function of your physical ability to gather
visual information and the way your brain manipulates it.
Your brain in this context is somewhat like a filter on a
camera lens. It can change the complete nature of the object
being photographed. Some filters will change just the color;
others will change the shape; yet others will create multiple
images or add special effects. The brain differs from a filter
in that the brain is a dynamic organ whose filtering capabilities are not static and are often based on personal experience
and expectation.

The fact that all visual information is processed by the brain
has some negative as well as positive aspects. The brain has
a tendency to interpret what it sees in light of what it expects
to see. If you are expecting to see a hill in a certain place or
land at an airport with a particular runway orientation, the
tendency is to try to fit what you see into your expectation
and ignore obvious discrepancies. The stronger your expectation, the easier it is for you to misinterpret reality.

There are many examples of expectations causing pilots to
misinterpret the available information, sometimes resulting
in tragic accidents. O'Hare and Roscoe (1990, page 40) relate
the case of Air New Zealand Flight Number 901 — a DC-10
that flew into Mount Erebus in Antarctica on a sight-seeing
flight:

> "It is possible to misinterpret observed features in
> line with the pilot's expectancy of what should be
> visible. The Air New Zealand DC-10 crash pro-
> vides a tragic example of this.
>
> "Unknown to the crew, the aircraft's programmed
> course had been shifted 25 miles to the east, in
> line with Mount Erebus. From the cockpit record-
> ings recovered after the crash, it is apparent that
> all the flightdeck crew members were attempting
> to monitor position visually by looking for fea-
> tures that would be visible if they were on course.
> Unfortunately, . . . the features actually visible
> were similar to those they expected to see and
> were easily misidentified, leading the pilots . . . to
> believe that they were well clear of the volcano.

"The circumstances of this particular accident have been presented with great thoroughness elsewhere. [See Recommended Readings.] **The tragic outcome of the Air New Zealand Flight 901 serves to underline the importance of an understanding of visual perception by the pilot. The apparent obviousness of what we perceive and our lack of conscious awareness of the complex cues involved in visual tasks combine to represent a hazard to the unwary and unprepared pilot.**" [Emphasis ours.]

It is easy to underestimate the ability of the brain to distort reality. The following dramatic example should leave you with no doubt as to how much this can happen. If you put on glasses with lenses that invert the images you see, initially, it is almost impossible to function because everything is upside down. However, after a few days (and a great deal of discomfort), the world returns to normal, even though you are still wearing the glasses. What this means is that the brain is initially confounded by the inverted images and cannot deal with them. After a while, it is able to take these images and invert them so you can function normally once again. Of course, when you remove the inverting glasses, the situation is repeated with everything being initially inverted followed by a reorientation back to what you are used to seeing.

It is not really surprising that this happens. Without any glasses at all, the images of what you look at are inverted on the retina, so your brain has already processed this visual information and inverted it to make life easier for you.

We have raised this example, not because it has any direct implication for flying (Do NOT try to fly with inverting glasses!!), but because it shows the extent to which the brain is involved in the process of seeing.

ILLUSIONS AT AIRPORTS

For a pilot, the area of vision that should cause the greatest concern is visual illusions. Unfortunately, there are a lot of normal situations that are easily misinterpreted due to the nature of the visual information received by the eye and processed by the brain.

Size Differences

Perhaps the most obvious example is when you are approaching to land at an unfamiliar airport. If you are used to flying at an airport that has an 8,000-foot (2,440 meters) runway that is 300 feet (90 meters) wide, you get used to it looking like runway A in figure 4-6, assuming you are on a normal three-degree approach path. If the unfamiliar airport has an 8,000-foot runway that is only 100 feet (30 meters) wide, it will look like runway B, if you are on a three-degree approach. Your reaction is likely to be that your approach is too high and that you should descend. If you followed your instincts, this would result in a dangerously low approach, especially if other visual illusions are also present.

Sloping Runways

A similar situation occurs when the unfamiliar airport has a runway that is not level. If you are on a perfect three-degree approach to a runway that has even a small uphill slope, you will perceive yourself to be too high. The tendency is to descend to make the approach look normal, bringing you too close to the ground. If the runway not only slopes up, but also is narrower than you are used to, the effect of being too high will be magnified, and you will tend to fly a dangerously low final leg.

Similarly, if the runway has a downhill slope, you will tend to think you are too low. This can cause you to fly too high an approach. While this is not intrinsically dangerous, it can lead to an overshoot of the runway. If you suddenly realize

A

B

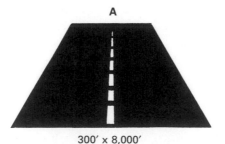

Figure 4-6   300' x 8,000'   100' x 8,000'

that you are too high and try to correct by pushing the nose down, you run the risk of trying to land at too high an airspeed.

The "Black-Hole" Effect

The "black-hole" effect is another illusion that has been cited as being a potential cause for several accidents. It occurs when a night approach is made over unlighted terrain, such as water. In this case, the black area in the foreground causes the bright runway lights in the distance to appear to be lower than they really are. This gives you the impression of being higher than you really are. An increase in the descent rate in this situation could be disastrous.

The "black-hole" effect has been cited as the reason for a number of airline accidents. In fact, the Boeing 727 was involved in a number of accidents of this type shortly after its introduction. This caused Boeing psychologist Conrad Kraft to investigate the issue. It was he who first alerted the aviation community to the phenomenon.

Figure 4-7 summarizes the visual illusions due to terrain effects that are associated with approaches. The type of approach that most often results from these illusions is also given.

APPROACH ILLUSIONS		
Situation	**Illusion**	**Result**
Upsloping Runway or Terrain	Greater Height	Lower Approaches
Narrower-Than-Usual Runway	Greater Height	Lower Approaches
Featureless Terrain	Greater Height	Lower Approaches
Downsloping Runway or Terrain	Less Height	Higher Approaches
Wider-Than-Usual Runway	Less Height	Higher Approaches
Bright Runway and Approach Lights	Less Distance	Higher Approaches

Figure 4-7

Countering Runway Illusions

The scary thing about these runway illusions is that they affect even experienced pilots. However, there are some procedures that can help prevent these problems. First, always fly a consistent traffic pattern. That is, try to fly the downwind leg at a constant pattern altitude (say, 1,000 feet). Second, try to turn base at the same relative position to the end of the runway. Third, initiate your descent at the same point and establish the pitch and power setting that produces a constant airspeed and rate of descent — we call this a stabilized approach. And fourth, use as many other cues as possible to supplement your vision, such as knowing the AGL height of your normal approach as you turn base or when you are half a mile from the end of the runway. If VASI lights are in operation, visual illusions are easily countered by keeping on the glideslope.

In any event, consider going around whenever you think your approach is not what you would like it to be. Anytime extraordinary piloting skills are needed to salvage a landing, it is best to go around and try again. Some pilots are reluctant to do this because they feel embarrassed, or they feel it demonstrates weakness or failure. In reality, the only person to think in these terms is the pilot. At a controlled airport, the tower personnel are only too happy to have an airplane go around. It is far better for them to sequence the airplane back into the pattern than to close the airport because of a damaged airplane on the runway. The same principle applies at an uncontrolled airport.

A good rule of thumb with respect to landing is to go around if you feel rushed to get everything in place for touchdown. It is in these types of situations that you are likely to forget something critical, such as extending the landing gear or watching for other traffic.

SEE AND AVOID

There are some other visual illusions that are worth discussing, particularly those that relate to potential mid-air collisions. There are many ways a mid-air collision can occur. However, there are three that are directly related to visual illusions — when airplanes are coming head-on, when they are converging from the side, and when they are climbing or descending on the same path.

First, let us consider the case where two airplanes are coming straight toward each other at the same altitude. You may wonder how this can happen, particularly if everyone follows the hemispherical rules. It is possible and legal for two planes to be at the same altitude while going in opposite directions if they are below 3,000 feet AGL, where the hemispherical rule does not apply. Even above 3,000 feet AGL, two airplanes can find themselves on a collision course. This can happen through poor altitude control, such as when a VFR flight going east and an IFR flight going west are both a couple of hundred feet off their designated altitudes. It can also happen when one airplane is climbing or descending through the other's altitude.

The visual illusion when two airplanes are approaching each other head-on is less a case of the brain misinterpreting visual information than it is of the information itself being confusing. In general, it is always easier to see something if it moves. If a distant airplane is coming straight toward you, it is difficult to see because there is minimal **relative motion**. That is, the airplane does not appear to move relative to the ground, clouds, or even you. If you hold this book at arm's length, figure 4-8 shows the approximate size of a Cessna 172 at different distances. For example, at one nautical mile

Approximate Distance		View	Approximate Time to Impact
1 n.m.	1850 m		14 seconds
1/2 n.m.	925 m		7 seconds
1/4 n.m.	460 m		4 seconds
1/8 n.m.	230 m		2 seconds
1/16 n.m.	115 m		1 second

Figure 4-8

(1,850 meters) a small plane will appear very small. From wingtip to wingtip, the airplane will take up only a quarter of a degree out of the full 360° horizon. This is like looking at a bottletop 15 feet (5 meters) away from you. If you do see the airplane, it is difficult to gauge how far away it is, because its size changes so slowly. At one-half mile (925 meters), the small airplane will appear only half a degree wide — a barely noticeable change.

The problem is that, when the airplane becomes easy to see, its size changes so rapidly it is almost too late to take evasive action. In figure 4-8, we also show the time to impact of two airplanes closing at a speed of 250 knots (463 km/hr.) — a typical closing speed of two general aviation airplanes. When the Cessna is a quarter mile (460 meters) away, it appears only one degree wide, which looks very small. At the closing speed of 250 knots, impact is only four seconds away — that's right, FOUR seconds. Not much time to react and then act!

Unfortunately, the situation is made worse by the natural response of people (and animals) when confronted by an object that looms suddenly in front of them. The normal reaction is to throw up your hands in protection, to try to move backwards, and to stare at the oncoming object, all of which makes constructive action difficult.

It is an interesting sidelight that this instinctive reaction is exploited by some animals such as the elephant. When threatened, the elephant spreads its ears to appear larger, trumpets, and makes an initial charge. If you are being charged, you will have an instinctive, fearful reaction that is extremely difficult to control. You will have the same reaction if an airplane suddenly looms in front of you, and you may not be able to respond constructively.

The second type of potential mid-air collision is when two airplanes are traveling in the same general direction, but are on a converging course. Normally, you detect other airplanes by their relative movement against the background of clouds or ground. However, if an airplane is traveling in more or less the same direction as you at the right speed, it does not appear to be moving relative to you. That is, it has no relative motion and is hard to see. [Figure 4-9]

This converging airplane is also subject to the same issues of visual angle as the one coming head-on. The visual angle changes so slowly the airplane does not appear to be coming any closer. The rate of change is significantly slower than for an airplane approaching head-on, because the rate of closing is less. So, it may take many seconds for the airplane to move from one mile to one-half mile away. When these changes are so slow, they are hard to detect and, even if detected, they are easily ignored. Once again, however, when the converging airplane suddenly starts to loom large, there may not be enough time to take evasive action.

A third type of potential mid-air collision exists when an airplane is climbing directly underneath another one going in the same direction. This is not really a visual illusion, but a case where your vision is obstructed. The pilot in the airplane that is climbing cannot see behind and above, while the pilot in the airplane above cannot see down. The only way to minimize the risk in these situations is to take some preventative steps. If you are cruising straight and level in the vicinity of an airport, or if you are climbing or descending on an airway, it is wise to make shallow S-turns to improve your ability to see other aircraft. This also helps make your airplane more visible to other pilots. In addition,

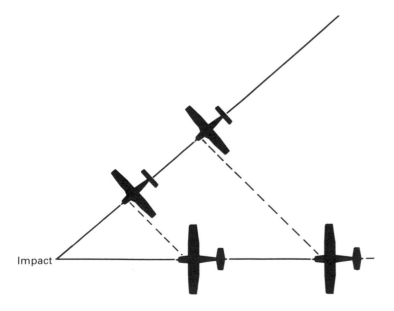

Impact

Figure 4-9

it is wise to look above you as much as possible, usually by leaning forward and looking upward through the windscreen.

AIR QUALITY

Another visual illusion we want to discuss occurs when you are flying in conditions of varying air clarity. As with almost everything else you do, you get used to making judgments about the world around you based on repetition and experience. In particular, you form habits of estimating how far away objects are by looking at them. For example, you can look at a runway you are familiar with and know with reasonable accuracy its distance from you. However, when you find yourself in situations that are different from normal, your ability to make accurate estimates decreases.

If you have ever flown in mountainous regions of the United States, you will know that mountains or towns that appear to be only a few miles away can often be much farther. Runway lights in New Mexico may look as though they are 15 miles (28 km) away but, in reality, they are 25 or 30 miles (46 to 56 km) away. The reason for this is that the air is crystal clear with little or no humidity so light passes through it with little absorption. The amount of reflected light reaching your eyes is the same as it would be from a much closer object in less clear air. This gives the impression of the object being close.

A similar illusion occurs when the air is misty or hazy, except that objects now appear farther away. This can have potentially dangerous consequences. For example, if you are approaching an airport in limited visibility, the runway would appear to be farther away than it really is. Obviously, this can result in you suddenly finding yourself having to rush your landing checklist as you approach the airport sooner than expected. Alternatively, you may end up being too high for the approach.

FIXATION, MOTION, AND FLICKER

Although optical illusions may be the most hazardous of the issues related to vision, there are several other conditions that warrant mention: namely, fixation and the effects of motion and flicker.

We have all experienced fixation in one form or another, not necessarily while flying. **Fixation** is the condition that occurs when you focus intently on one object to the exclusion of everything else. It is not strictly a visual condition, but is closely related to attention. It occurs most often, however, when the visual system is involved.

A typical example of fixation in instrument flying is concentrating on returning your airplane to its assigned altitude by looking at the rate-of-climb indicator and the altimeter. In this case, it is very easy to ignore all the other instruments and wander off your heading. Even more dangerous is a situation where your attention is diverted from your primary task of flying the airplane, such as when you pore over a map trying to locate an airport or VOR frequency.

The Eastern Airlines crash into the Everglades, in which the crew was intent on replacing an indicator bulb, has been partly blamed on fixation. The crew was so intent on the task at hand that the normal warning indications were either not heard and seen or ignored. The result was that the airplane flew into the ground. The lesson to be learned from this is that you should always keep scanning the instruments if IFR or looking outside if VFR. That is, your primary attention should always be on flying the aircraft, not focused on rectifying some aspect of the flight.

Vision and its relationship to pilot performance can also be affected when our vision is partially obscured in some way by natural or man-made moving objects.

One of the most dangerous examples of this is when you fly into rain or snow. The movement of the precipitation onto and around the windshield can be distracting, making it difficult to focus on objects outside the cockpit. In fact, this type of situation induces a type of empty-field myopia, as well as a hypnotic trance in which it is difficult to concentrate. Often, this effect is made worse by the regular and monotonous motion of windshield wipers. In such cases, you should be very careful to avoid staring at the rain, snow, or wipers, especially when transitioning from instruments to visual on an instrument approach.

Another dangerous situation arises in the presence of a flickering light. A flickering light can have a strong adverse effect on some people, occasionally causing nausea and sickness. This can happen even when there is no motion of the airplane. For example, you may start a single-engine airplane while facing the sun. When the engine is at idle and the propeller is spinning quite slowly, the flicker caused by the sun passing through the rotating blades can cause disorientation and nausea. If this affects you, move your airplane as soon as possible so the sun is not behind the propeller. The same thing can happen when descending into the sun at idle power.

A similar effect occurs when flying in haze or clouds with strobe-type, anti-collision lights on. The regular, repeated flashing reflected into the cockpit can cause the same effects as the rotating propeller. Any adverse effect on how you feel, of course, will have a negative impact on your performance.

RECOMMENDATIONS FOR OVERCOMING VISUAL ILLUSIONS

As we have seen, vision is intimately related to performance for several reasons. The eyes are the primary source of information, so any real or perceived distortions of reality will affect your ability to fly well.

In this chapter, we have tried to show how complex the visual system is in its own right, and how interrelated the eyes and brain are. Implicitly, we have been trying to provide some guidelines to help you avoid some of the hazards we have mentioned. Of these guidelines, three are most important:

1. The more aware you are of the limitations of the visual system and the problems associated with it, the better prepared you will be to avoid the problems.

2. Always use multiple sources for your information. Do not rely on a single source, even if it seems to be providing correct information.

3. Never ignore data that are in conflict with your expectations. That is, don't try to make the data fit a conclusion. Rather, use data to form conclusions.

THE EARS

We now turn our discussion to the other major way you receive data: namely, through the ears. For the most part, this comprises information from various air traffic controllers, such as the tower. It also includes listening to identifying signals from VOR and NDB transmitters, conversation with crew and passengers in the airplane, the wind, and sounds from the engine. Unfortunately, it also includes the inevitable noises associated with each of these sources of information. And, finally, the ears are intrinsically involved in determining your physical orientation with respect to the earth.

EXPECTATION

Earlier in the chapter, we discussed problems in vision caused by expectation. If you expect to see something, your brain distorts what the eyes see to satisfy that expectation. The same thing can happen with sounds, although aural illusions are less common.

A typical example of the effects of brain distortion of sound occurs when a controller issues an instruction that differs from what you expect. If the standard pattern at your airport is to the left, and the tower instructs you to enter right traffic, you may well hear the instruction as "left." Your brain was expecting the usual instruction and misinterpreted the new one to conform to the expectation.

This type of problem can be reduced if you are aware of its existence and pay attention to the words the controllers are saying, rather than assuming they are saying what you want them to say. The best solution, of course, is to repeat back what you understood the message to be. This practice also provides the controller with feedback that you have both heard the message and understood it.

ORIENTATION

It goes without saying that knowing which way is up is critical in flying. On the ground, you generally have no problem doing this unless you are excessively intoxicated or dizzy from medication, illness, or spinning. In the air, it is much more difficult for you to use your senses to tell what is happening to you in relationship to the earth's surface. It is very easy to feel that you are in one position or orientation when, in fact, you are in a very different one.

Most problems associated with orientation occur when the visibility is limited, where the eyes alone cannot provide sufficient information by looking outside the cockpit for the brain to know which way is up. So, the brain seeks information from elsewhere to come up with the solution. In this section, we discuss how the brain does this.

We start with a cautionary note. Disorientation can happen to anyone, even instrument-rated pilots. However, noninstrument-rated pilots are much more prone to it. Disorientation is also one of the most common reasons why pilots have accidents. It is not only very unsettling, but extremely dangerous to lose orientation when flying. It is also very easy. In fact, it is impossible to maintain orientation when flying using the body's sensory organs alone, unless you can see the horizon.

VESTIBULAR SYSTEM

The body relies on several sensors to tell it which way is up and what is happening to it in terms of motion. The **vestibular system** in the inner ear provides balance and movement information. It is comprised of three semi-circular canals filled with fluid, the vestibular sacs which detect acceleration, and the vestibular nerve. When the eyes and vestibular systems provide corroborative information, the brain functions well. When the two systems provide conflicting information, the brain has enormous difficulty processing the information, often resulting in acute loss of orientation and motion sickness.

The three semi-circular canals are orientated at right angles to each other. They lie in three planes that correspond approximately to those of an airplane. [Figure 4-10] Each canal is filled with fluid and has hair-like fibers that sense the motion of the fluid within the canals. When your body is at rest, the fibers stand straight up, because the fluid and canal are stationary. If the canal and fluid are traveling at the same speed; that is, in constant motion with no acceleration, the fibers also stand up.

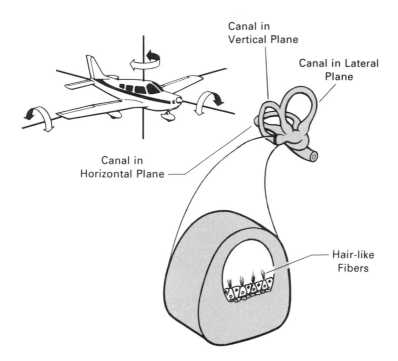

Figure 4-10

Consider the case when the canal and fluid are stationary. If the head now turns, the canal turns with the head, but the fluid's motion lags, causing the fibers to lean away from the turn. You can see this for yourself if you attach a short thread to the inside of a glass. Fill the glass with water and let it stand for a while. The thread will hang straight down. If you now rotate the glass counter-clockwise, the water's inertia prevents the fluid from moving as rapidly as the glass, and the thread leans away from the rotation of the glass. [Figure 4-11]

This same principle applies to the inner ear. It provides rapid information about how your body is moving; however, it is at this stage that one of the problems occurs. If the glass continues to turn at a constant rate, the water eventually catches up with the glass, so that both are rotating at the same rate. And what happens to the thread? Because the glass and water are no longer moving differently, the thread hangs straight down.

When this happens in the semi-circular canals and the fibers no longer lean to one side or the other, the sensory system

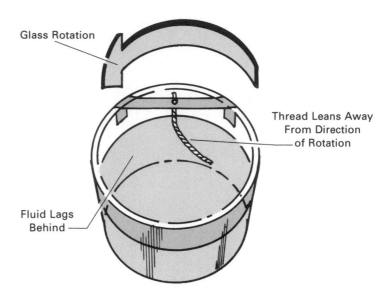

Glass Rotation

Thread Leans Away
From Direction
of Rotation

Fluid Lags
Behind

Figure 4-11

believes the body is again at rest, even though it may actually be turning. That is, the vestibular system has provided erroneous information to the brain.

Unfortunately, the situation can become worse. If the rotating glass is briefly decelerated, the fluid's inertia keeps the water moving at the original speed. Once again, there is a discrepancy between the speed of the glass and the speed of the fluid, and the thread will now lean to the right **in the direction of the rotation**. [Figure 4-12] In fact, the thread leans in the direction opposite to the prevailing acceleration.

Let us consider what happens in your head when this situation occurs. Initially, assume your head and the fluid in the canals are rotating clockwise at the same rate. This implies the fibers are vertical. If the body is decelerated, the fluid will continue to rotate at its original rate while the canals slow down. The fibers now lean to the right away from the deceleration, which is counter-clockwise. The brain interprets this to mean the head is now rotating counter-clockwise. However, the head is still rotating clockwise, which is its original direction, so the brain has been fooled once again.

It is difficult to believe that it is so easy to fool your brain. To become a believer, you should take part in a simple

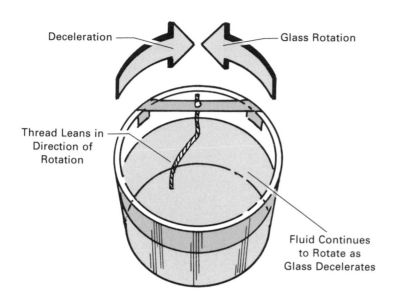

Deceleration — Glass Rotation

Thread Leans in Direction of Rotation

Fluid Continues to Rotate as Glass Decelerates

Figure 4-12

demonstration that illustrates it. Find any smoothly rotating chair or bar stool — the smoother the better. Blindfold a friend and have him or her sit in the chair with the head in a normal upright position facing straight forward. Instruct them to remain seated, even if they think the chair has stopped moving. Now, spin the chair so it is rotating at a constant rate. Have your friend report verbally what is happening.

Initially, the friend will say that the chair is being rotated, which is accurate. After several seconds, your friend will report that the chair has stopped rotating. This is inaccurate, because the chair is still moving at a constant speed. Now, touch the chair to decelerate it slightly. Your friend will say the chair is moving in the opposite direction, even though it is still rotating in its original direction.

When you watch your friend, you may find it difficult to believe the reports. It is very strange to listen to information that so clearly contradicts what you are seeing. To check for yourself, exchange places with your friend and you will see that the reports were all accurate. The FAA often gives this demonstration at its seminars, and we recommend you attend one.

The demonstration can be taken one step further. This time have your blindfolded friend sit with his or her head tilted to one side and facing down — the same position that you would have if you had to lean over to change fuel tanks or pick up something on the floor. Once again, spin the chair. When the inner-ear fluids have had a chance to stabilize, stop the chair abruptly and have your friend sit upright. Your friend will most probably jump off the chair — so be prepared!

What is happening here is that the fluid in all the canals starts rotating, because the head is tilted at an angle to the rotation. When the chair stops rotating and the head is placed in an upright position, the fibers sense motion in all canals. This signals the brain that the body is falling to one side, causing it to compensate by leaning or jumping in the opposite direction. Watching this happen in an experiment is startling to say the least.

The relevance of this to flying is obvious. Your body cannot be relied upon to provide accurate information with respect to orientation, particularly in limited visibility. In several studies, private pilots who flew into clouds almost all ended up losing control of the aircraft, many within 30 seconds. This happened because they relied on their senses for orientation rather than the instruments in the airplane. Had these pilots been by themselves, and not with an instructor, most would have crashed.

The Leans

One of the most common sensations in flying when visibility is restricted is known as **the leans**. This occurs when the airplane has slowly entered a turn without you noticing it (a common situation). That is, even though the plane is now banking, your brain senses it to be in straight-and-level flight. When you notice the bank, the general tendency is to rectify it quickly, returning to straight-and-level flight. This rapid corrective action creates motion of the fluids in the inner ear, signaling the brain that your body has tilted. Because the brain initially thought it was level, the tendency is for you to lean in the direction of the original turn.

In an extreme case, the leans can lead to disastrous consequences. For example, if the airplane is in a prolonged spin or graveyard spiral, the inner ears stabilize even though rotation continues. On pulling out of the spin or spiral, you experience the sensation of spinning in the opposite direction. This can lead to a reaction that will put the airplane into another spin in the same direction as the original one.

RULES OF THUMB

Disorientation is a complex topic, but an important one, because we are all prone to it. Some types of disorientation are easily noticed because of a mismatch of cues. Others are insidious, because no mismatch is available or noticed.

There are some rules of thumb that are worth following:

1. Do not trust your senses.
2. Believe the instruments.
3. Be aware of situations that are likely to increase disorientation. Common among these are stress, being distracted, flying under the influence of alcohol or medication, flying with a cold, and moving your head rapidly away from an upright position. In addition, be alert for disorientation when operating in reduced visibility such as low ceilings, haze, fog, or at night.

Disorientation is a major cause of accidents, especially when pilots who are not competent to fly on instruments find themselves in the clouds. The best remedy, of course, is prevention. That is, not to fly into a situation that requires more skill than you have. We also recommend that you practice instrument flying. Three or four hours with a competent instructor might be enough to provide the basic skills necessary to get out of the clouds if you inadvertently fly into them. These skills need to be maintained. If you have not flown on instruments for some time, take a refresher flight or two either in a simulator or an airplane.

CONCLUSION

In this chapter, we have discussed the body's two major sources of information: the eyes and ears. Because so much of the information you need to fly comes from these two organs, it is important to understand how they work and how well the data they sense is transmitted to the brain.

It is the unfortunate truth that both our eyes and our ears can be sources of misinformation as well as facts. What this means for you, the pilot, is that you have to examine the quality of information you receive to ensure it has not been distorted in some way, and you should always try to corroborate the information from a second source.

Exercises

Chapter Questions

1. Draw a picture similar to figure 4-4 that shows what happens to someone who is farsighted. Then, add the corrective lens for farsightedness.

2. What effect would being without corrective lenses have on a shortsighted pilot when landing?

3. Draw a diagram of the eye showing the arrangement of the rods and cones and the position of the fovea.

4. Describe the problems you may encounter when flying in low light. Why do these occur?

5. Describe the phenomenon known as empty-field myopia. When is it likely to occur and what can you do to reduce its effects during flight?

6. Describe how differences in runway width and slope can affect the perception of your approach.

7. Describe three types of potential mid-air collisions and the reasons why they occur.

8. What is visual fixation? Give an example of it in the cockpit.

9. Describe the vestibular system and how it works.

10. Explain why it is unsafe to trust your sense of orientation when flying in clouds.

11. Describe "the leans."

REFERENCES AND RECOMMENDED READINGS

Dehnin, G., Sharp, G.R., & Ernsting, J. 1978. *Aviation Medicine*. London: Tri-Med Books.

Gregory, R.L. November 1968. "Visual illusions." *Scientific American*.

Gregory, R.L. 1978. *Eye and brain*. New York: McGraw Hill.

Hawkins F.H. 1987. *Human Factors in Flight*. Aldershot, UK: Gower Technical Press.

Kraft, C.L. 1978. "A psychophysical contribution to air safety: Simulator studies of visual illusions in night visual approaches." In H. Pick, H.W. Leibowitz, J.R. Singer, A. Steinschneider, and H.W. Stevenson (Eds.). *Psychology from research to practice*. New York: Plenum.

Leibowitz, H.W. 1988. "The human senses in flight." In E.L. Wiener and D.C. Nagle (Eds.). *Human Factors in Aviation*. San Diego, CA: Academic Press.

Mahon, P. 1984. *Verdict on Erebus*. Auckland, New Zealand: William Collins.

O'Hare, D. & S. Roscoe. 1990. *Flightdeck Performance: The Human Factor*. Ames: Iowa State University Press.

Vette, G. 1983. *Impact Erebus*. Auckland, New Zealand: Hodder and Stoughton.

Wiener, E.L. & D.C. Nagel (Eds.). 1988. *Human Factors in Aviation*. San Diego, CA: Academic Press.

The Brain 5

INTRODUCTION

The human brain is one of the most complex mechanisms known. It is the central control and information processing center for almost all human activity, as well the storage place for information received over the course of a lifetime. No computer yet devised has the brain's processing or memory capacity. It can retrieve primary information, such as a face not seen in 20 years, faster than the fastest computer. Brain functions are so complex that scientists will be discovering more about them for centuries to come.

The brain plays a central role in aviation human factors. It receives information about the flying environment, processes that information, makes decisions about what to do, and then signals the hands, feet, and speech organs to respond. It is also the point where emotions, stresses, and non flight-related pressures are applied to the pilot decision-making process. Fortunately, we do not need to understand the brain completely to make effective use of its functions.

In this chapter, we describe the basic parts of the brain and how they function for storing, retrieving, and processing information. We also discuss some of the brain's attributes that have a direct impact on your ability to fly safely.

THE MAJOR COMPONENTS OF THE BRAIN

The brain has four major components. [Figure 5-1] First, the **cerebellum** is a small, double-shell-shaped section located just behind the top of the spinal cord. This part of the brain controls balance and muscular coordination, which are essential to the motor skill activities in flight that are sometimes called the "stick and rudder skills." The cerebellum is considered the most primitive part of the brain because it is fairly well developed in most lower-level animals as well as in humans.

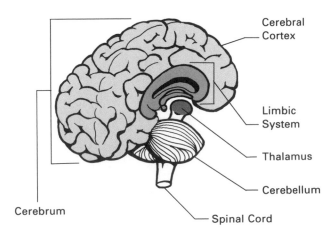

Cerebral Cortex

Limbic System

Thalamus

Cerebellum

Cerebrum

Spinal Cord

Figure 5-1

Second, the **thalamus** is a slightly smaller, egg-shaped part located above the cerebellum. It serves as a central control mechanism for messages going to and from the sense organs, such as the eyes, ears, and nose. Close to the thalamus is the hypothalamus (not shown), a small, round component that controls several body functions, including temperature, metabolism (the chemical processes of converting air and food into substances the body can use), and endocrine balance (hormone levels that affect energy, mood, and reaction to stress, among other things).

Third, the **limbic system** is a series of canals located around the thalamus that controls emotions and sequential activities, such as carrying out a planned series of activities.

Fourth, the **cerebrum** is the largest part of the human brain. It fills the area directly under the skull and surrounds the other parts of the brain. Here, all complex mental activity such as thinking, remembering, sensory perception, and voluntary movement is initiated. The outer surface of the cerebrum, called the cerebral cortex, controls the various sensory organs (the eyes and ears) and motor activities (walking, running, grasping, and so on).

As shown in figure 5-2, the cerebrum consists of two parts, called hemispheres. Each hemisphere is connected underneath by a heavy series of nerve connections, called the

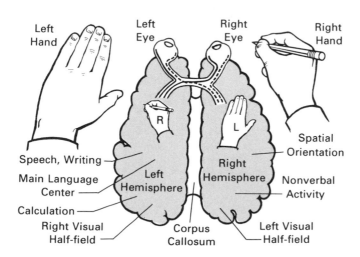

Figure 5-2

corpus callosum. Both hemispheres have some areas that control similar functions such as sensory, motor, visual, and auditory areas. However, the left side of the brain receives visual information from the right side of the visual field, and the right side of the brain receives the left side of the visual field. The same can be said for most other left/right activities. Similarly, each hemisphere controls parts of the body on the opposite side.

Some brain functions are unique to one side. The left side of the brain, which is dominant in most people, performs most verbal activity, such as written and spoken communication, as well as analytical calculation. The right side performs most of the non-verbal activities, such as spatial orientation, music, art, dance, and fantasy. Both sides are, obviously, crucial to flying.

The brain is the core of the **central nervous system** (CNS). There are millions of nerve connections running throughout the body, which are essential for communication between the organs and the brain. The organs send information about their status to the brain, and the brain responds with commands on how to react. Many of the nerve connections run to the brain through the spinal cord, which is sometimes considered an extension of the brain. In skilled, psychomotor performance, such as acrobatic maneuvering ("seat-of-the-pants" flying), the signals probably go directly from the spinal cord to your responders (your hands, feet, and so on) without going to your brain for conscious processing.

BRAIN FUNCTIONS

When information is received from your eyes, ears, and other sensors, it is converted into electrical impulses that travel through nerve fibers to the brain. Your brain encodes the signal according to its content, context, and time-relationships. The information is then stored. The information can later be retrieved from storage by searching for the context in which it was stored. This whole process is very complex and beyond the scope of this book. However, certain parts of the process have an impact on your performance as a pilot.

MEMORY

There are two types of memory: short term and long term. **Short-term memory** is the memory you have of events and information received in the very recent past, usually no more than the last thirty seconds. It has a small capacity (usually about seven unrelated items) and a high rate of decay. That is, unless it is rehearsed or used, information in short-term memory disappears quite rapidly. For example, when the tower gives you a complex clearance, you are unlikely to be able to remember it all without writing it down. If you rely on your memory, the chances are good that you will forget or distort some of the information given to you.

On the other hand, **long-term memory** is considered permanent storage. Information that is rehearsed (practiced) or is of primary importance may be encoded and placed in permanent storage in long-term memory, where the capacity is much greater and the decay rate is slower. Information stored in long-term memory and not used frequently may also disappear over time — "use it or lose it" as the saying goes. This process is thought to be, in part, related to the amount of information added to memory in the interim. However, there is evidence to support the notion that once information reaches long-term memory, it is never lost. It just becomes more difficult to recall. In studies where parts of the brain have been electrically stimulated, long-forgotten events have been recalled vividly. See our chapter on learning to fly for a more complete discussion of the conditions under which information is best learned.

FORGETTING

At one time or another, everyone forgets something that they want to remember. It is likely that every pilot has forgotten to do something crucial to safe flight. Often, the results of these lapses in memory will never be forgotten again. For example, the second author will never forget the time, on his second attempt to pass his commercial checkride, that he forgot to replace the oil dipstick during the preflight. A most embarrassing film of oil appeared on the windscreen shortly after takeoff with the examiner on board.

Scientists have three explanations for why people forget: decay through time, interference, and motivated forgetting. It is likely that all three can act simultaneously and may reinforce each other.

DECAY THROUGH TIME

Some scientists think that much forgetting is simply a result of the passage of time. You are sure to recall times when you crammed for an examination only to forget the material soon thereafter. The FAA recognizes the fact you are likely to forget certain things about flying and requires you through regulation to practice flying periodically. Three takeoffs and landings every 90 days and biennial flight reviews are examples of regulations that help restore forgotten skills and procedures.

Some flight tasks are not forgotten as quickly as others. For example, continuous hand-eye tasks, such as manipulating the flight controls, are not easily forgotten, even though they can become rusty. This is similar to the saying, "You never forget how to ride a bicycle." On the other hand, you are likely to quickly forget procedures, like starting the airplane, working the audio panel, or extending the landing gear manually. The explanation is that continuous eye-hand manipulation tasks become over-learned and automatic and can be done without thinking. Procedures, on the other hand, are usually not repeated as frequently and, thus, are not learned as well. Therefore, procedures are forgotten more quickly.

INTERFERENCE

Interference is the process whereby new information being added to memory tends to interfere with the recall of information already there. This theory is supported by research that shows information can be recalled much better if sleep is the only activity between learning and recall, as compared with being awake and experiencing other mental stimulation after learning.

In flying, you perform many procedures that interfere with each other, causing you to forget some of them. Research shows that the more similar two procedures are to each other, the more likely they are to interfere with each other. For example, you are more likely to forget or confuse the procedures for short- and soft-field landings (which are similar) than those for a short-field and an emergency landing (which are dissimilar).

MOTIVATED FORGETTING

In some cases, you will forget something because you want to forget it. Your brain has a sympathetic way of storing information that is unpleasant, such as the events surrounding an accident or death, in a way that is not easy to recall. If you were able to recall extremely painful memories easily, it could cause you great mental anguish. It is as if your mind has a built-in mechanism to protect you from memories that would be damaging.

IMPROVING YOUR MEMORY

Almost everyone could benefit from an improved memory. On average, once a day in the USA, one pilot forgets to lower the landing gear before touching down. There are many circumstances in flying in which you are expected to remember a lot of information that is coming to you rapidly. For example, when flying VFR from a busy airport, you are often given detailed take-off and departure clearances. You are also expected to be prepared for an immediate takeoff, which means you must remember these clearances, because you do not have time to write them down.

Many techniques are available to assist your memory. Some are designed to improve your memory, while others provide an alternative source for the information. Of the memory-improvement techniques, the most common are the use of rehearsal or practice, creating meaningful structures of the information, and using mental images. Of the supporting techniques, writing in shorthand and use of checklists are the most common.

REHEARSAL

Rehearsing what you want to remember is usually done by repeating the information out loud either to yourself or to someone else. This is the primary way that actors remember their lines. Another form of this technique is to teach the information to someone else. As we mentioned in an earlier chapter, many teachers have said that they did not really know the material until they had to teach it. So, teaching someone else the material or procedures you want to remember can be a very effective technique for learning it well.

ORGANIZING THE MATERIAL

Another technique for helping memory is to organize the information into patterns or chunks that make sense. Since you are usually limited to remembering about seven pieces of information in your short-term memory, chunking or organizing the information can dramatically increase the amount you can remember. For example, if you are told that the wind is from 270 degrees at 15 knots, it is usually easier to remember the numbers "twenty-seven, fifteen" than the complete statement. When you need the information, you recall the numbers and decode them into their components.

IMAGERY

Another way to help improve your memory is to associate a mental image with each item of information you would like to remember. For names, you can associate some particular feature about the person to be remembered. For mundane information, you can think of visual objects, such as a dog or cow and mentally associate it with the information. This association technique can be enhanced if you look at pictures together with the information. In aviation, such visual images are quite easy to generate because of the dependance you have on seeing things.

MNEMONICS

Perhaps the most common technique in aviation for improving memory is through use of mnemonics (pronounced nee-MAHN-iks). These are mental tricks that facilitate memory, such as GUMP (Gas, Undercarriage, Mixture, Props), BUMPF (Brakes, Undercarriage, Mixture, Pitch, Fuel), or TMFFISHH (pronounced TM-fish) for fixed-gear airplanes (Throttle, Mixture, Fuel, Flaps, Instruments, Speed, Harness, Hatch). Mnemonics provide both a catchy word to remember and an organization of the information. The problem with mnemonics is that you have to rely on your memory to activate them and to work out what each of the letters means.

SHORTHAND

Using shorthand to write down information as it becomes available in the cockpit, such as in an amended IFR clearance, is very helpful because it bypasses your memory. It provides a hard-copy of the information that you can refer back to at any time. The success of the technique depends on your ability to read the shorthand after you have written it. We recommend that you either use a standard shorthand or develop a consistent one of your own. If you make it up as you go along, you will have to rely on your memory to figure out what you wrote. The problem with using shorthand is that you need a firm surface to write on, and you need a free hand to write with; this can be troublesome in periods of high activity.

CHECKLISTS

Of all of the techniques, the one we favor the most is the use of checklists, even in a single-pilot environment. We prefer checklists because the items you have to do are in front of you, and you do not have to rely so much on memory. Obviously, you still have to remember to use the checklist and you have to use it properly. But, this is more likely to occur than remembering every item on the list. As with shorthand, using checklists can be cumbersome during busy times.

We also recommend forming good procedural habits, such as verbally acknowledging each item on the checklist, even to yourself. If one of the items on the checklist is "Mixture rich," say out loud, "Mixture rich?" and confirm it when done with the response "Mixture rich." You are more likely to make an error if you do this mentally rather than verbally. Some pilots have the habit of saying out loud "Gear indicates down and locked" as a matter of course in every landing, whether flying a fixed-gear or retractable airplane. This voicing of the monitoring action can actually cause you to check that the gear is down rather than simply thinking you have checked it.

No matter which technique you use, you should constantly check and double-check everything you do. This is particularly true for procedures that are typically not part of checklists or that are separated by time. For example, even if setting cowl flaps is part of a climb checklist, it is common to forget to set them upon reaching altitude, because your attention may be focused on communicating with ATC and you last used the checklist several minutes earlier.

Efficient use of your short-term memory is essential to flight operations, especially for communications and short-term planning. In particular, pilots in IFR or busy VFR conditions are expected to be able to hear, understand, and respond to rapid voice commands from air traffic control. It is essential that you develop your own individual technique for efficient and accurate information storage and rapid recall.

THINKING

The thought process is a highly complex human activity that is not well understood. At the primitive level, the thought process is concerned with dealing with the physical environment. As humans have developed, they have added abstract symbols in the form of oral and written language to make it easier to work and live together. For example, if you are in the forest and hear a lion roar, your mind generates an image of a lion. To communicate those thoughts to another person, you use abstract symbols of the form LION to communicate it in writing, and abstract sounds to communicate it aurally. In both cases, the other person's mind must transform the written or aural message into the image of a lion before he or she can respond appropriately to your communication.

In aviation, the same process holds true. For example, on the downwind leg, you read the altimeter and form an image of your height above ground. You also form images representing what you should do to adjust your altitude if it is not what it should be.

There are some thought processes in aviation that do not require mental images. For example, the concept of time and the computation of estimated time of arrival require no mental images associated with physical objects. These types of mental processes are done entirely in abstract form, represented by our language and system of numbers. However, we are still principally physical beings and process concrete information faster than we do abstract.

It has been well established, as a principle in human factors, that we can make decisions faster and more accurately if we are presented with concrete, real-world forms with which to process than if we are presented with abstract forms, such as numbers or words. Abstract forms require a transformation before they can be used to make a decision. For example, if you were shown your distance from a VOR-defined course in the form of a number representing miles, you would have to transform that number first into direction from the course and then into magnitude (how much change of heading is needed to get back to course). It can be done, but it is harder to do than if your position is presented as a moving needle. You could process the moving information even more easily if it were presented on a pictorial map display so that you could see the whole picture.

INFORMATION PROCESSING

Models of the human information processing system are presented elsewhere in this book. At this stage, we want to discuss one limitation of the system that affects your ability to think about and do more than one thing at a time. Many scientists believe that we have only one channel through which all thoughts must go, and this becomes the limiting factor in the ability to focus attention on more than one thing at a time. Of course, you know that people can walk and chew gum at the same time. However, these are motor activities that are controlled separately from the cognitive thinking process. Physically, these automatic processes are probably controlled in the spinal cord. On the other hand, conscious information processing is controlled by higher systems in the cerebrum, and it appears that humans are limited to one of these at a time.

How does this limitation affect us in aviation? For the most part, it means that you are limited in the number of decisions you can make within short intervals of time. For example, you may be concerned about the buildup of ice on the wings and, at the same time, be faced with a communications breakdown. Your brain requires that your thought process be done in series, one after another, which means that you cannot solve both problems simultaneously. It also means that you are better off prioritizing and spending slightly longer on each problem than switching rapidly from one to the other. Of course, you should never devote all your time and concentration to the solution of a single problem, but should at least monitor the status of the other aspects of your flight. In most cases, this is a dynamic process with changing priorities.

Mental workload is directly associated with the limitations described above. There are certain times during a flight when workload tends to peak. The greatest predictable workload tends to occur during the takeoff and final approach phases of flight, as well as during the initial climb-out phases. During these times, any additional work, such as an ATC request for a change in your flight plan, is very difficult to handle. The reason is that you can only process one thing at a time, and your ability to divide your attention between many tasks is poor.

ATTENTION

Attention can be defined as the mental faculty that controls the subject matter chosen for conscious information processing. Not much is known about how this mechanism works, but it is probably closely associated with motivation, which is discussed below.

How you control your attention is a very important element in flying, because accidents are often caused by the attention of the pilots being diverted from the primary flying task. For example, when an Eastern L-1011 descended into the Florida Everglades, the NTSB cited the crew's diversion of attention to a burned-out light bulb as a causal factor.

The attention mechanism can be seen in what has been called, "the cocktail party phenomenon." If you are in a large group of people, you can listen to one conversation while filtering out other conversations around you. Your attention mechanism chooses the person to listen to and you can hear that person very well in spite of the other voices. You do not even hear the other conversations because of the limitation discussed above of being able to pay attention to only one thing at a time.

However, your capacity is not entirely limited. For example, if your name is mentioned in another conversation at the same sound level or even lower, you will probably hear it and divert your attention to that conversation. This phenomenon seems to indicate that your channel for conscious processing may have a continuous additional capacity for processing certain simple, but important bits of information.

In aviation, this additional information processing capacity is both a blessing and a curse. It is a blessing because it can call your attention to items that are important to you while you are thinking about something else. For instance, as soon as you hear your call sign, you will stop a conversation in the cockpit and pay close attention to a radio transmission. It can be a curse because it can divert your attention from rational, safety-oriented thoughts about a needed decision, to irrational background-oriented thoughts, such as getting home. Through training and experience, you can learn to discipline your attention mechanisms so you attend to items that are safety oriented.

MOTIVATION

Motivation is a force coming from within your brain that causes you to act or behave in certain ways. It is usually considered a positive force because it causes you to act or move forward as opposed to remaining stagnant. Motivation manifests itself either in intensity or in direction. We say that a person is motivated if he or she is taking action about something. Motivation does not give any positive or negative societal value to the form of action being undertaken. In other words, you can be motivated to do what society considers either wrong or right. In aviation, you can be motivated to take risks (for the attention you may receive) or to make safe decisions (for the feeling of security and sense of responsibility you feel).

Historically, the concept of motivation has undergone several changes depending on the orientation of thinking about human psychology at the time. During the time of the "rationalists" in the 19th century, all human activity was attributed to rational thought. There was no room for motivation with these philosophers, because behavior was the result of rational choice. Later, in the 20th century, psychologists described motivation in terms of "needs" or "drives." People have certain drives, such as hunger, thirst, sex, or achievement, and these motivate people to behave the way they do.

Both of these viewpoints are useful in explaining much of what motivation is about. However, neither completely describes the motivation force. We know that the presence of a stimulus object such as food can bring on an appetite.

Therefore, it is not just the need or drive that motivates, but also the presence of stimulus objects. In the chapter on judgment, we will discuss how both rational thought and motivation play a part in pilot decision making.

Within the brain, it is not entirely clear where motivation is triggered. However, there appears to be a strong center for its regulation in the hypothalamus. Researchers have found that electrical stimulation of certain areas of this organ can cause strong drives for food and drink.

EXPECTANCY

In every situation in which you become involved, whether on the ground or in flight, you develop an awareness of the predictable elements of the situation. From this awareness, you learn to expect certain things to happen. This becomes like a mind set which reduces the amount of things you have to think about because you say to yourself, "this situation has occurred before, so I can expect the same result again."

For example, suppose you call ATC and request 9,000 feet on an eastbound IFR flight. ATC returns the call saying "Cleared to 10,000 feet." You have in your mind set an expectation of 9,000 feet for two reasons. First, you requested it and you usually get what you ask for from ATC; and, second, it is an appropriate altitude for the direction you are flying. In this example, if there is anything restricting the transfer of communication, such as static, partial blocking, or your attention focused on something in the cockpit, you may not hear (that is, understand) the message as 10,000 feet because you are expecting 9,000 feet.

There are many examples of such expectations that have led to accidents. The KLM and Pan Am 747s at Tenerife that collided on the runway is one such example. The tower cleared the KLM into takeoff position to hold. The captain expected to hear cleared for takeoff and the transmission was partially blocked. He was so sure that he had received the takeoff clearance that he did not question the tower further, even in the face of some uncertainty expressed by his copilot. This combination led him to make the takeoff when the Pan Am airplane was still on the active runway, resulting in the worst air disaster up to that time.

CONCLUSION

We have seen in this chapter that the human brain is an extremely complex organ. Scientists are far from fully understanding its functions. On the other hand, there are many functions of the brain that are fairly well understood which are highly important to flying performance. These functions include thinking, memory, forgetting, recall, and psychomotor performance. Our purpose in this chapter was to give you a brief background in these functions to help you understand yourself better because such awareness can improve your flying performance.

Exercises

Chapter Questions

1. List the four basic parts of the brain and describe the function of each.

2. What are the two types of memory and how do they function?

3. Describe some of the limitations of the brain that can lead to errors in performance.

4. Briefly describe the various theories of why people forget.

5. What are some things you can do to improve your memory?

6. Why are checklists a good idea? What are their drawbacks?

7. What is expectancy and how does it affect your thought process?

8. Describe a situation in which expectancy caused you to do something incorrectly. What happened and why? It need not be within aviation.

REFERENCES AND RECOMMENDED READINGS

Atkinson, R.C. & R.M. Shiffrin. August 1971. "The control of short-term memory." *Scientific American.*

Hilgard, Atkinson, R.C. & Atkinson. 1975. *Introduction to Psychology.* New York: Harcourt, Brace, Jovanovich.

Hubel, D.H. September 1979. "The brain." *Scientific American.*

The Body 6

INTRODUCTION

In the previous chapters, we introduced how aircraft design can influence your effectiveness; how the eyes and ears function and provide information to the brain; and how the brain processes this information, makes decisions, and acts on the data it receives. In this chapter, we introduce how the body influences your ability to function under the various conditions encountered in flight. We will look at five major areas: breathing; food and drugs, and how the body reacts to what we put into it; environmental factors, such as noise and vibration; fatigue and stress; and the role exercise plays to keep our bodies in peak flying condition.

BREATHING

Breathing, and its vital organ the lung, is at the core of your existence. If you stop breathing, you die, because the brain becomes starved of oxygen. In this chapter, however, it is not the extreme case of respiratory failure that we are most interested in; it is the situation where breathing is impaired by one cause or another.

In very simple terms, the act of breathing involves the transfer of oxygen in the lungs to the bloodstream. This is called **respiration**. As you inhale, oxygen in your lungs fills very small air sacs, called alveoli. The high pressure in your lungs then forces the air from the alveoli into your bloodstream, where it is carried by hemaglobin in the blood to the various organs needing the oxygen. This process is called **transportation**. Anything that reduces the ability of the lungs to transfer oxygen or the ability of the blood to carry it causes the body to function less than optimally.

As the cells in your body function, they give off carbon dioxide as a natural waste product. This is carried by the bloodstream back to the lungs where it is passed to the alveoli and expelled from your body as you exhale. The whole cycle from inhaling to exhaling normally takes place at a rate of 12 to 15 times a minute. Under normal circumstances, this rate is controlled by sensors in the lungs that monitor the amount of carbon dioxide. If there is too little carbon dioxide, your breathing rate increases, and vice versa. There are also sensors that monitor the level of oxygen in the blood. As this falls, the sensors instruct the brain to increase the respiration rate. Finally, you can also consciously increase your rate of respiration if the circumstances call for it, but you should be careful because this practice can lead to hyperventilation, which we discuss later.

We start our discussion by examining what it is we breathe: namely, air.

THE AIR

The earth is surrounded by layers of gases, particulate matter, and water vapor. We call this mixture **air**. All the layers are often referred to collectively as the **atmosphere**, which is held around the earth by gravity.

Air comprises many different constituents, the two most important of which are nitrogen (about 78%) and oxygen (about 21%). The remainder includes traces of argon, carbon dioxide, helium, and hydrogen, to name a few; as well as variable amounts of beneficial gases, such as water vapor and ozone; and harmful gases, like carbon monoxide, sulfur dioxide, nitric oxide, and other oxides that cause acid rain. There are also minute proportions of dirt and dust.

The atmosphere is divided into layers, each with different characteristics. As a general aviation pilot, you will rarely venture out of the first layer: namely, the **troposphere**. It extends upward from the earth's surface to between 20,000 and 60,000 feet (6,100 and 18,290 meters), depending on a variety of conditions.

The troposphere can be variable and unstable with respect to temperature and water content; that is, it is not easy to predict either of these in advance without extensive weather data. The only aspect that is generally true is that temperature decreases as altitude increases. At the upper edge of the troposphere there is a thin layer called the tropopause, which acts as a lid to confine most of the water vapor.

The atmosphere is most dense at the earth's surface with density decreasing the farther one moves from the surface. That is, there is more air per unit volume at the surface of the earth than there is high above it. This is the reason it requires more runway to take off at high-altitude airports.

One way to determine air density is by measuring the pressure of the atmosphere; that is, the weight of the air per unit area. Instead of doing this directly in units of pounds per square inch or kilograms per square centimeter, it is customary to measure pressure in terms of how tall a column of mercury the atmosphere can support.

This convenient method is standardized worldwide; however, the unit of measurement is not. In English units (now only used in the United States!), the measure is expressed in inches of mercury. Elsewhere, it is expressed in millimeters of mercury or, more commonly, in millibars, also sometimes called hectopascals. In most countries, the millibar is the standard unit of pressure measurement in aviation.

ALTITUDE		PRESSURE		
Feet	**Meters**	**Inches Hg**	**Millibars**	**Millimeters Hg**
0	0	29.92	1013.2	760.0
3,000	910	26.82	908.2	681.2
6,000	1,830	23.98	812.1	609.1
9,000	2,740	21.39	724.3	543.3
12,000	3,660	19.04	644.6	483.6
15,000	4,570	16.89	572.1	429.0
18,000	5,490	14.95	506.3	379.7
21,000	6,400	13.19	446.8	335.0
24,000	7,320	11.61	393.2	294.9

Figure 6-1

To understand how quickly air pressure (and hence density) decreases with altitude, look at the chart in figure 6-1. At about 18,000 feet (about 5,490 meters) the air is half as dense as it is at sea level.

From a physiological perspective, the atmosphere can be divided into three layers which we call the safe, or physiological, layer; the dangerous, or physiologically deficient, layer; and the impossible, or space equivalent, layer. [Figure 6-2] In the **safe layer**, which extends up to about 10,000 feet (3,050 meters), humans live freely. In the **dangerous layer,** which extends from 10,000 to about 50,000 feet (3,050 to 15,240 meters), humans usually need support in the form of oxygen equipment, although adaptation is possible (such as in the mountains of Tibet). Humans cannot survive in the **impossible layer** without complete and all-encompassing support (oxygen, heat, and so on). This section of the chapter deals mainly with the dangerous layer.

HYPOXIA

A pilot's performance is impaired anytime the body does not properly transfer oxygen into the bloodstream or transport it efficiently through the body. **Hypoxia** is the term used to describe the body's reaction whenever it is deprived of the amount of oxygen it needs.

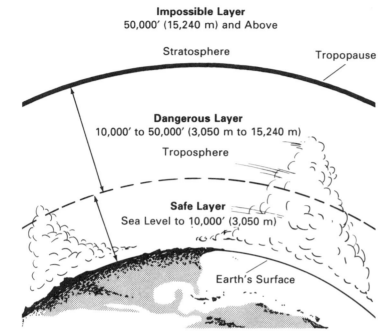

Impossible Layer
50,000' (15,240 m) and Above

Stratosphere   Tropopause

Dangerous Layer
10,000' to 50,000' (3,050 m to 15,240 m)

Troposphere

Safe Layer
Sea Level to 10,000' (3,050 m)

Earth's Surface

Figure 6-2

There are different ways such deprivation can occur. Hence, there are different forms of hypoxia. When there is insufficient oxygen in the lungs for proper transfer to take place, the condition is called **hypoxic** hypoxia. When the blood's capacity to carry oxygen is impaired, it is called **anemic** hypoxia. When the flow of blood is hindered, we call this **stagnant** hypoxia. And when the body has a reduced capacity to accept oxygen from the blood, we call this **histotoxic** hypoxia. In all these cases, the net effect is the same — reduced oxygen to the body, most importantly to the brain and eyes.

HYPOXIC HYPOXIA

Hypoxic hypoxia occurs anytime the body cannot obtain enough oxygen from the lungs to service its needs. Drowning is an extreme form of this type of hypoxia. Hypoxic hypoxia is also caused when the lungs do not transfer the available oxygen efficiently. For example, any form of lung disease diminishes respiration, as does smoking (more on smoking later).

From a pilot's perspective, however, the most common cause is being at too high an altitude without supplemental oxygen. As we mentioned earlier, air density decreases at higher altitudes. Thus, there is less oxygen to breathe and less to transfer into the bloodstream. Below 10,000 feet, the effects are negligible for most people. Above this altitude, however, the effects become increasingly dangerous.

ANEMIC HYPOXIA

Anemic hypoxia occurs when there is sufficient oxygen in the lungs, but the blood has lost some of its capacity to carry it. This can occur from a lack of blood cells, such as exists if you suffer from anemia or after donating blood. It also exists when the cells do not combine at the normal rate with oxygen, as occurs in carbon monoxide poisoning, the most common cause of anemic hypoxia in aviation.

Carbon monoxide finds its way into the air mainly from engine emissions and smoke from tobacco products, such as cigarettes and pipes. Carbon monoxide is more than 200 times more attractive to the hemoglobin in the blood than oxygen. When carbon monoxide is present in inhaled air, a disproportionate amount of carboxyhemoglobin is formed causing an oxygen deficiency throughout the body. In small amounts, this causes impairment of brain functions and sight; in larger amounts, it causes death.

This form of hypoxia, then, occurs when there are exhaust gas emissions in the cockpit or when there is a smoker in the cabin. It should be stressed that tobacco smoke also affects nonsmokers.

STAGNANT HYPOXIA

For most general aviation pilots, this form of hypoxia rarely occurs. Stagnant hypoxia refers to the situation whereby the appropriate amount of oxygen is in the blood, but the blood pools and circulation is below normal. Typical examples are artery diseases such as arteriosclerosis and atherosclerosis, or varicose veins. Any heart failure or degradation also causes stagnant hypoxia. In aviation, this condition can also be brought on by the high G-forces present in acrobatic flight.

HISTOTOXIC HYPOXIA

The final form of hypoxia is caused by the body's inability to take the oxygen it needs from the blood. Although the blood itself may have the proper amount of oxygen and may be circulating efficiently, the body's cells have lost their ability to extract the necessary oxygen. This happens primarily when you are under the influence of alcohol or drugs. It would also happen if you were suffering from cyanide poisoning.

SYMPTOMS OF HYPOXIA

Hypoxia is a major problem in aviation because it is an extremely insidious condition. That is, it creeps up on you, and usually gives a feeling of well-being in the process. So, not only does it impair your ability to perform, but since it makes you feel good at the same time, there is often a dramatic discrepancy between how you perceive your own performance and how external observers assess it.

The table in figure 6-3 shows just how big this discrepancy can be. In the left-hand column is a list of symptoms as seen

SYMPTOMS OF HYPOXIA	
Observer	**Person with Hypoxia**
Mental:	**Mental:**
Euphoria	Euphoria
Poor performance	Good performance
Confusion	
Impaired judgment	
Physical:	**Physical:**
Increased respiration	Dizziness
Poor coordination	Nausea
Unconsciousness	Headaches
	Tingling
Behavioral:	
Aggressiveness	
	Vision:
	Blurring
	Tunnel vision

Figure 6-3

by an objective observer. In the right-hand column is a list as felt by a person suffering from hypoxia.

TIME OF USEFUL CONSCIOUSNESS

One of the most common measures of the effects of hypoxia is the Time of Useful Consciousness (TUC). This is the time you have to continue to perform tasks in a reasonably competent manner without supplemental oxygen. As altitude increases, your ability to perform declines rapidly. [Figure 6-4] At 18,000 feet (5,490 meters), you have only a short TUC (20 to 30 minutes). At 25,000 feet (7,620 meters), this time reduces to a mere 2 to 3 minutes.

The effect of hypoxia varies considerably by individual; however, there are some factors that make the effect worse. They include the rate at which you ascend — the faster you ascend the greater the effect; whether or not you are physically active — the more active the shorter the TUC; whether you are physically fit — the better condition you are in, the longer it will take for hypoxic effects to show; and smoking — smokers almost always have a shorter TUC. Some studies have shown that for smokers an altitude of 5,000 feet (1,520 meters) can be equivalent to an altitude of 10,000 feet (3,050 meters) for a nonsmoker.

TIME OF USEFUL CONSCIOUSNESS			
Altitude		**While Sitting Quietly**	**During Moderate Activity**
Feet	Meters		
40,000 Ft.	12,200	30 Sec.	18 Sec.
35,000 Ft.	10,670	45 Sec.	30 Sec.
30,000 Ft.	9,140	1 Min. and 15 Sec.	45 Sec.
25,000 Ft.	7,620	3 Min.	2 Min.
22,000 Ft.	6,710	10 Min.	5 Min.
20,000 Ft.	6,100	12 Min.	5 Min.
18,000 Ft.	5,490	30 Min.	20 Min.

Figure 6-4

REGULATIONS

In recognition of the potentially disastrous effects of hypoxia on a pilot's performance, there are regulations governing the altitudes at which you are allowed to fly without supplemental oxygen.

In the United States, you may fly as a required crewmember for no more than 30 minutes above a cabin pressure altitude of 12,500 feet, and you may never fly above 14,000 feet, unless supplemental oxygen is used. These regulations recognize that even at lower altitudes, such as 12,500 - 14,000 feet, hypoxia begins, especially if some time passes. Above 14,000 feet, the effects of hypoxia can be so severe that performance will be severely impaired.

We recommend that you always adhere to these regulations because, by its very nature, you may not notice the onset of hypoxia. We also recommend that you consider setting lower personal limits if you are a smoker or prone to the effects of hypoxia.

ALTITUDE CHAMBER

In order to experience hypoxia firsthand in a safe environment, we highly recommend that you take a ride in an altitude chamber if you ever have the chance. They are not easy to find, but the FAA in Oklahoma City, some military bases, and the University of North Dakota in Grand Forks, North Dakota, have well equipped facilities. You will be amazed at the experience.

During a chamber ride, you will be asked to perform simple tasks, such as arithmetic, and you will believe you answered properly. Yet, after the ride, when you look at the results of your efforts, you may see just a scrawl. Another common task is to put differently shaped pegs into their respective holes in a board. At 25,000 feet (7,620 meters), it is a remarkable sight to see people struggling with this task.

As you can tell from this discussion, we regard hypoxia as a major problem for general aviation pilots because it is so hard to monitor and detect. Once you notice the effects, it may be too late to correct the problem. In addition, if you begin to suffer from hypoxia by being at too high an altitude, your respiratory system will demand more oxygen causing you to breathe faster. This, in turn, can lead to hyperventilation, which compounds the problem.

HYPERVENTILATION

Hyperventilation is the term used to describe breathing that is too fast and too deep, resulting in an excess of oxygen and a deficiency of carbon dioxide, which is one of the triggers for breathing. Hyperventilation usually occurs involuntarily as a response to a stressful situation and has symptoms very similar to those of hypoxia. Fortunately, it is easily remedied by a return to a normal breathing pattern. This can be done either by consciously lowering your breathing rate or by conducting an ordinary conversation. Breathing slowly in and out of a paper bag also helps because it raises the carbon dioxide level in your lungs.

FOOD AND DRUGS

In the first section of this chapter, we dealt with breathing and the various ways it can impair performance. In this section, we discuss the effect of substances that we ingest into the body, substances that are both healthy and harmful.

DIET

It is generally accepted that people perform better if they eat a balanced diet. A good diet provides the body with the proper nutrients to build and maintain its cells, to regulate a variety of bodily functions, and to supply energy. A bad diet, on the other hand, can create a set of circumstances that will adversely affect your performance as a pilot.

HYPOGLYCEMIA

The most noticeable effect of a poor diet is **hypoglycemia**, which is low blood sugar. This condition exists naturally when you wake up in the morning. With a proper diet, the sugar level will rise to a level in keeping with the needs of your body.

If you eat food with a high refined-sugar content (donuts, soft drinks, or candy) early in the morning, your sugar level will rise rapidly to a very high level. This results in the "high" so often associated with the intake of quick-energy foods. It is the reason athletes often eat a candy bar just before playing. Unfortunately, the rapid high is usually followed by a rapid drop in blood sugar to levels that have a profound physiological impact. This drop results in tiredness, headaches, and an inability to concentrate. In extreme cases, which are fortunately rare, unconsciousness can result. This condition is caused by the excessive secretion of insulin into the blood stream as the body strives to rid itself of the excess sugar.

The lesson to be learned here is to prepare for a flight by eating well beforehand. In this way, your body will not become hypoglycemic and, hence, will not crave a sugar fix. During a flight, do not consume refined sugar in order to keep awake.

CAFFEINE

Another substance people often use to keep awake is caffeine. It is found in coffee and many drinks and, frequently, it is combined with sugar. Caffeine is a stimulant that can increase alertness and reduce reaction times; however, its usefulness varies from one individual to another.

Its use as a stimulant is controversial. Some people regard it as a drug and some religions expressly forbid its use. Others regard it as a relatively harmless substance that causes no problems if used moderately. Caffeine certainly dehydrates the body to some extent, which can be detrimental over a long period of time, but this should not be a concern with only a cup or two of coffee.

ALCOHOL

Alcohol, on the other hand, is a major problem in a demanding profession such as flying. Because of the detrimental effects of alcohol on the body and mind, Federal Aviation Regulations restrict its use before and during flight.

PHYSIOLOGICAL EFFECTS

Alcohol disrupts almost all body functions necessary for safe flight. It impairs the visual and auditory systems. It has a negative effect on both short-term and long-term memory, as well as on the processes of thinking and decision-making. It also slows the reflexes and makes coordinated movement more difficult.

Perhaps worst of all, it warps how you perceive yourself. It makes you reckless, while at the same time leaving you with the impression that you are performing well. In this way, its effects are very much like those of hypoxia, where judgment and decision-making are degraded, while self-confidence is heightened.

All these effects are present even if you drink socially; a mere one or two drinks will have a negative impact on your ability to fly safely. If you drink heavily, the effects can last for days. Finally, the effects of alcohol are stronger at greater altitudes.

In general aviation in the United States, at least 10% of the fatal accidents between 1975 and 1981 were related to alcohol. The actual number of accidents, fatal or not, related to alcohol is most probably higher because no statistics are available on the alcohol content of pilots who survive.

REGULATIONS

In the United States, the regulations regarding drinking and flying are very clear. There are two parts to the regulations: you may not drink within eight hours before you fly; and you may not fly while under the influence of alcohol, defined by having .04 percent or more by weight of alcohol in the blood. In light of the potential prolonged effect of excessive drinking, the first part of the regulation can be too lenient. The second part, however, is designed to take care of such situations.

From our perspective, flying while under the influence of alcohol, no matter how little, is extremely dangerous. As mentioned above, alcohol makes you feel that your performance is improving even though an objective observer would be able to see otherwise.

DRUGS

There are two types of drugs — those used for "mind-alteration," such as LSD, marijuana, and cocaine; and those used for medical reasons, which are usually purchased over-the-counter or prescribed by a physician, such as tranquilizers, sedatives, and amphetamines.

The very nature of drugs that create an altered sense of being implies that they are not congruent with flying. Aviation is extremely demanding of all your facilities. Anything that alters the ability of these facilities to function optimally increases the likelihood of an accident. We believe it is potentially suicidal to fly if you are taking any of these kinds of drugs.

The other types of drugs are those prescribed by doctors or that are available over-the-counter. Typical are tranquilizers and sedatives designed to relieve stress, and amphetamines to work against the effects of sleepiness. There are also drugs for allergies, colds, and numerous other ailments.

There are two things you should consider before flying while using drugs. First, what is the condition you are treating; and, second, what are the side effects of the drug or drugs used to treat the condition? Some conditions are serious enough to prohibit flying, even if the illness is being treated successfully with drugs. Taking a drug is an indication that you are not well. Your main concern should be whether the illness itself is severe enough to affect flight safety. In many cases, it is.

It is also important to realize that many doctors prescribe medication without knowing you are a pilot. Always let your physician or pharmacist know you are a pilot and ask about the side effects of medication. You should also consult an aviation medical examiner (AME) about any medication you suspect will adversely affect your ability to fly safely.

Unless specifically prescribed by a physician, do not take more than one drug at a time and, especially, do not mix drugs with alcohol. The effects are often unpredictable. Many doctors recommend that you not take any alcoholic beverage within 24 hours of medication, and that you not fly within 24 hours of taking medication. Although this may be an extreme precaution, it does highlight how seriously the potential effects are viewed.

The ability of the body to function properly is directly affected by what you ingest. On the one hand, being careful of what you eat, drink, and inhale can have beneficial effects on your body and its ability to perform. On the other hand, abusing your body can have adverse effects on it and its ability to perform.

Abuse of your body through the use of drugs or alcohol causes an increased susceptibility to hypoxia, slower reaction times, lowered concentration, and impaired judgment. With a list like this, the conclusion should be obvious. Look after your body at all times, and never fly under the influence of substances that will impair performance.

ENVIRONMENTAL CONSIDERATIONS

In this section, we briefly discuss several aspects of your flying environment that may impact performance. We have already discussed the effects of exhaust fumes on the body, particularly carbon monoxide. Now let us focus on other environmental factors such as ambient temperature, noise, and vibration.

NOISE

Of the environmental factors affecting the general aviation pilot, noise deserves the most attention. Noise can have detrimental effects on communication, both within the cockpit and with air traffic control personnel. Obviously, this can have a negative affect on how well you fly.

Noise can also have permanent, damaging physiological effects. How you hear is partly related to the loudness of the sound, which is measured in decibels (dB). A whisper will be in the 30-40 dB range, while a shout can be as loud as 100-110 dB. The typical cockpit of a small aircraft maintains a level of about 70-90 dB. Prolonged exposure to sounds exceeding about 80 decibels can damage the ear and reduce your ability to hear properly. In extreme cases, even a single very loud noise, such as an explosion, can damage or destroy hearing.

The adult human ear is capable of hearing sounds that have frequencies as low as 20 Hz to as high as 20,000 Hz. Hz stands for Hertz which is the measure of frequency in beats per second. As a reference, middle C on a piano is 274 Hz. When the ear is damaged, it is the ability to hear high-frequency sounds that typically becomes impaired. We recommend regular checks of your hearing, particularly if you fly frequently or are exposed to extremely loud noises.

Both the frequency and loudness of a sound affect our ability to hear it. Our ears are more sensitive to high frequency sounds, so if two sounds are equally loud, the one with the higher frequency will be perceived as louder and will more likely attract our attention. This is why alarms are almost always high-pitched.

Similarly, some people speak so softly that it is very difficult to hear what they say. In the presence of ambient noise, such as is found in the cockpit of a small airplane, hearing them can be even harder than normal. In terms of intelligibility, the ratio of the spoken voice to the surrounding noise is an important factor in whether the voice will be heard. The louder the voice is in relation to the surroundings, the easier it will be to hear. However, there is a confounding factor. As both the voice and the noise become louder, even in the same ratio, it becomes increasingly difficult to distinguish the voice.

Noise in the cockpit of a small airplane can also cause fatigue and stress. You are most susceptible when the noise is loud or high-pitched, such as the hiss of the wind passing around the fuselage. Both fatigue and stress can impair your ability to fly well and are discussed later in this chapter.

We have a single recommendation to make with respect to noise. On a day-to-day basis, you should take every precaution to mask out noise by wearing earplugs or a well-fitting headset. If you do this, you will remain more alert during the flight and reduce the chances of long-term hearing damage. However, there is one drawback of using headsets or earplugs with embedded speakers. If you switch to a radio which has its volume high, there is a chance that you will damage your eardrums.

Our preference is for simple earplugs, even inexpensive ones made from sponge-like materials. These provide excellent protection without any side-effects. Unfortunately, a lot of pilots do not like to wear ear protection because they think it will affect their ability to hear radio communications. This is not true. Earplugs will lower the intensity of both the voices you want to hear and the background noise, and since the ratio of the two remains constant, you will still be able to hear the voice. In many cases, wearing ear protection can actually improve intelligibility.

VIBRATION

Another environmental factor for general aviation pilots to consider is vibration, although it does not usually have a major impact. Prolonged exposure to vibration has an effect similar to noise: it induces fatigue and stress. Depending on intensity and location, it can also cause headaches and muscular discomfort, all of which may serve to distract your attention. There is not a great deal you can do about vibration, except be aware of its possible effects. Vibration found in flying rarely causes permanent damage to your body.

TEMPERATURE

The final environmental factor for our consideration is temperature. Although it is rare for it to have a significant impact on performance while flying, it is useful to know about the effects of temperature if you ever find yourself in an abnormal situation.

The human body is most comfortable when operating in a temperature range of about 70° to 80° Fahrenheit (21° to 27° Celsius) with a relative humidity of approximately 50%. When the ambient temperature strays too far from this, the body becomes stressed by being either too hot or too cold.

The first type of heat stress is likely to occur when your operating environment is too hot, defined as in excess of 90° F (32° C). The most noticeable effect will be tiredness, and it will become increasingly difficult for you to concentrate. You may also suffer from heat exhaustion, which occurs when your circulatory system cannot compensate for blood vessels that have dilated to increase heat loss. Drinking water is a good antidote for heat stress.

A second type of heat stress that can affect your performance can occur if you have had too much sun prior to a flight, through sun bathing or being outside without adequate protection. It can also occur when you have not had sufficient liquid intake. The effects may be subtle and more difficult to notice, because your environment may be comfortable and you may feel well. However, your concentration may be lowered, leading to a decline in your decision-making ability.

A third type of heat stress can occur when the temperature falls below about 50° F (10° C). If the temperature is very low, it is possible for your body to lose more heat than it can produce. In extreme cases, this will cause shivering (a heat producer), drowsiness, and poor concentration. All of these will have an adverse effect on your performance.

The best way to deal with potential problems from extreme temperatures is to be sensible. If it is very hot, drink a lot of water and keep your skin protected. If it is very cold, make sure you are warmly clothed and do not rely on the heating system of the aircraft. If it fails, you could be in serious trouble. If you are flying in extreme temperatures, you should also have an appropriate survival kit on board in case of an emergency landing or crash. In summer, plastic bottles of water are essential. In winter, you should have warm clothes, matches, and high-energy food. Alcohol is not recommended because it gives a false sense of warmth while actually lowering your resistance.

FATIGUE

There are two types of fatigue that affect a pilot. The first, called **acute** fatigue, is caused by intense mental or physical activity at a single task. Examples of this are having to work extremely hard on an important problem while under pressure, or flying in instrument conditions in turbulence for four or five hours. Anytime you have to give a task your undivided attention for a prolonged period, you are likely to suffer from acute fatigue.

The second type of fatigue is **chronic** fatigue. It is caused over time by such factors as lack of sleep, jet lag, and stress. Acute fatigue can be cured by rest, such as a good night's sleep. However, chronic fatigue may take much longer to disappear, and can only be alleviated if the root causes are eliminated.

Fatigue affects every part your performance. It degrades attention and concentration resulting in missed cues or information; it affects decision making; it has a negative impact on coordination; and it makes you less able to deal with other people, particularly in situations of conflict.

Fatigue also tends to focus your attention on things that require activity rather than thought and lowers your ability to handle the multiple tasks associated with flying. Furthermore, it lowers your discipline, encouraging you to accept greater margins of error or risk than you would if alert. When fatigued, it is difficult to muster the enthusiasm and energy to pay attention to detail.

Our recommendations in this regard are simple. Avoid flying when fatigued. Always try to have a full night's sleep before you fly, and do not fly after an exhausting day at work. Give yourself time to relax before a flight and time to prepare for it mentally.

If you become fatigued during a flight, land and rest. This is often a difficult decision to make, because sometimes the growing fatigue forces you to concentrate so hard on the flying that you are not consciously aware of how tired you are. Furthermore, once a journey has started, it is much more difficult to interrupt it for something as seemingly intangible as tiredness.

Fatigue is frequently linked with pilot error and you should never underestimate its detrimental effects.

STRESS

There are two major types of stress: physical and emotional. Of the two, emotional stress has the greatest impact on flying, and we discuss it in detail later in the book. However, physical stress is also detrimental to performance.

Physical stress usually occurs when you are exposed to prolonged physical discomfort, although some of the effects of emotional stress can also be physical. For example, if you fly without earplugs in a plane that has loud ambient noise, you will almost certainly suffer from physical stress. You will become fatigued quickly and may feel increasing tension in your neck and shoulder muscles. Similarly, if you have to fly for any length of time through turbulence, the constant buffeting will induce physical tension.

The best cure for stress is, of course, prevention. However, sometimes this is not possible, such as when flying through turbulence. If you feel physical stress coming on, you can reduce it by doing some simple exercises. Isometric exercises are possible, even in the cramped confines of a general aviation airplane. First, tightening and relaxing the neck and shoulder muscles is helpful, as is rolling your shoulders backward and forward. You could also have a passenger massage these muscles.

Second, deep breathing is beneficial, especially when you breath in with your stomach muscles. Breathing in with your stomach muscles results in your stomach moving outward not inward, as happens when you puff out your chest. This type of breathing pulls air into the depths of your lungs.

Finally, ensure that your body temperature is comfortable. As we have already mentioned, excessive heat or cold is, in itself, stressful and frequently impairs performance. In conjunction with other stress, it can exacerbate the situation.

Although physical stress is usually not as harmful to performance as emotional stress, it can still detract from your performance. Consequently, be aware of its effects and try to prevent its onset.

EXERCISE

Most of this chapter has dealt with ways that your body can mislead you, or with conditions that can cause it to malfunction or function poorly. In this section, we take a more positive approach and discuss some things that you can do to improve your performance as a pilot.

In addition to maintaining your skill and knowledge of flying through practice and study, the most beneficial activity you can do is exercise. Although experimental evidence is sometimes tenuous as to the direct links between exercise and performance, most physicians and psychologists agree that regular exercise is helpful.

It appears that regular, well-balanced exercise has positive effects in at least three ways. First, better muscle and aerobic conditioning leads to an increased ability to tolerate or counter fatigue. That is, a fit person can perform at a higher level for longer than a person who is not fit. It appears that this holds true both for physical and mental activities.

It is easy to imagine the growing physical discomfort of someone who has been flying for three or four hours without a break. If this pilot has poor muscle tone, it is likely that his or her posture will be adversely affected causing further discomfort. In addition, if the flight has been turbulent, the stresses of countering it will cause fatigue more quickly in someone who is not physically fit.

The overall effect of this increasing physical fatigue is a lowered ability to concentrate mentally. Of course, lowered concentration leads to degraded performance. In general, it appears that people who exercise regularly are able to concentrate for longer periods of time and to recover from fatigue more quickly than people who do not exercise.

A second positive effect of exercise is improved self-esteem. Exercise has been shown to increase self-confidence which, in turn, leads to a more positive attitude when flying. This is linked to the compelling evidence that regular exercise promotes a better emotional state. When people feel good about themselves and are confident, they usually are more motivated about what they do and deal with stress better. As you will read later in the book, increased motivation and lowered stress both lead to improved performance.

The third beneficial effect of exercise is more global in nature. There is no doubt that regular exercise improves overall health. A sedentary body is more prone to being overweight, which increases the chances of having cardiovascular problems, as well as unbalanced forces on the spine.

An active body has a better conditioned heart resulting in enhanced flow of blood and, hence, efficient oxygen transportation throughout the body. A person who is in good physical condition is also less likely to have a heart attack. Furthermore, exercise changes the body's metabolism, which allows active people to eat more with less propensity for gaining weight. Other benefits of exercise occur later in life, such as lowered chances of having fragile bones (particularly if the exercise includes some impact, such as aerobics, racquet sports, or running) and improved memory.

We have no doubt that your performance as a pilot will benefit from a good exercise program. This should typically include activities that exercise the heart and lungs, and ones that increase strength and mobility. If you are not exercising regularly, consult your physician about the type of program to undertake. You will be delighted at the results.

CONCLUSION

In this chapter, we have discussed how our bodies can affect our performance. We have dealt with breathing and hypoxia and with the effects of substances you ingest. All of these can adversely affect your ability to perform well. We also suggested that proper care of your body is important and strongly recommended a regular exercise program.

The health of your body has a direct impact on the ability of your brain and muscles to perform well. Consequently, taking care of your body will help improve your overall ability to fly well.

Exercises

Chapter Questions

1. Briefly describe the process of breathing.

2. Define hypoxia. Why is it such a problem?

3. Describe the four types of hypoxia and what causes them.

4.   From an observer's perspective, what are the symptoms of hypoxia?

5.   Describe the physiological effects of alcohol.

6.   Describe the differences between acute and chronic fatigue. How would you go about recovering from each type?

7.   What is hypoglycemia? How does it occur? What are its effects on pilot performance?

8.   Why is it important to let your physician or pharmacist know you are a pilot?

9.   Describe the short-term and long-term effects of high ambient levels of noise in the cockpit.

10.   What causes physical stress? What are its effects on your ability to fly safely?

11.   Why is regular exercise beneficial?

REFERENCES AND RECOMMENDED READINGS

Bryan, Stonecipher, and Aron. 1954. *University of Illinois Bulletin*. Champaign-Urbana, IL:

Dehnin, G., Sharp, G.R., & Ernsting, J. (Eds.), 1978. *Aviation medicine*. London: Tri-Med Books.

Gregory, R. L. 1978. *Eye and brain*. New York: McGraw Hill.

Hawkins, F. H. 1987. *Human Factors in Flight*. Aldershot, U.K.: Gower.

National Transportation Safety Board. 1984. *Safety Study: Statistical review of alcohol-involved aviation accidents.* Report NTSB/SS-84/03. Springfield, VA: National technical Information Service.

Parker, E. S. & E. P. Noble. 1977. Alcohol consumption and cognitive functioning in social drinkers. *Journal of Studies on Alcohol*, 38, 1224-1232.

Reinhart, R. O. 1982. *The Pilot's Manual of Medical Certification and Health Maintenance.* Osceola, WI: Specialty Press Publishers.

Ryback, R. S. & P. J. Dowd. 1970. After effects of various alcoholic beverages on position nystagmus and coriolis acceleration. *Aviation Medicine*, 41, 429-435.

Emotional Stress **7**

INTRODUCTION

Emotions are an enormously powerful set of forces in human nature. They can take us from the heights of ecstasy to the depths of despair in a short period of time. Emotions are the guiding forces behind almost all human activity ranging from art and music to war. When you fly, emotions can have a tremendous influence on your ability to perform well, particularly when it comes to issues of judgment.

For most people, flying tends to be an emotional experience. In the beginning of their aviation careers, people feel a great deal of joy and exhilaration as well as some fear as they "break the surly bonds of earth." Later, they feel privileged or powerful for having the capability of flight. Being able to fly sets them apart from other people and can make them feel very good about themselves.

At the same time, many pilots like to hide their emotions, almost as though acknowledging emotion makes them less capable or qualified. In reality, your emotions are always present, influencing your thinking and decision making. If you ignore them, you may not recognize the major impact emotions have on your ability to fly safely.

We think the discussion of emotions and the stresses they cause is of particular importance in general aviation, because most flights have only a single pilot. When this is the case, you, as a single pilot, have all the responsibilities for both making and implementing decisions. There is no copilot to assist you if you perform poorly.

In this chapter, our discussion focuses more on the results of emotions, particularly stress, than on the emotions themselves. We do this because we are more likely to succeed in teaching you to recognize stress and to take steps to counter its negative influence, than to change the underlying emotions causing the stress. Throughout the chapter, we provide practical information on techniques and exercises that can help you reduce the effects of stress. Some of the examples and illustrations in this chapter are taken from the FAA book entitled, *Aeronautical Decision Making for Student and Private Pilots* (May 1987).

STRESS

The term stress originates from engineering where it refers to the force placed upon an object to cause straining, bending, or breaking. In the human context, **stress** is commonly used to describe the body's responses to demands placed upon it, whether these demands are pleasant or unpleasant. Anything that causes stress is called a **stressor**.

Consider the following scenario:

> On a cross-country flight, you may suddenly realize that you are much lower on fuel than you expected. The clouds ahead appear to be building, and there is considerable static on the radio. You are off course and you cannot seem to find a familiar ground reference point. On top of this, you failed to take a comfort stop before the flight and now have a full bladder. The cabin heater is not functioning properly, and you are starting to encounter turbulence.
>
> You now have many things on your mind. You begin to worry about arriving late at your destination and missing an important appointment. You begin to worry about having to make a forced landing and damaging the aircraft, which a friend was not keen to lend you in the first place.
>
> Your palms are becoming sweaty and your heart is starting to pound. You feel a growing tension, and your thinking is becoming confused and unfocused. You give too much attention to the "what if" questions rather than dealing with the situation.

At this point, you are suffering from three types of stress: physical, physiological, and emotional. **Physical** stressors include environmental conditions, such as temperature and humidity extremes, noise, vibration, and lack of oxygen. **Physiological** stressors include fatigue, lack of physical fitness, sleep loss, missed meals that have led to a low blood-sugar level, discomfort associated with a full bladder or bowel, and disease. We discussed physical and physiological stressors in the chapter on the body.

Emotional stressors include the social or emotional factors related to living and intellectual activities, such as solving difficult problems in flight. It is this class of stressor that we emphasize in this chapter.

It is important to realize that the process of making a simple decision is one of the leading causes of stress. Even the simple commitment to make a flight can be very stressful, whether or not there is pressure being exerted on you by others.

The relationship between stress and performance has been verified in numerous experiments and is illustrated in figure 7-1. When stress is virtually nonexistent, motivation and attention are minimal and, as a consequence, your performance is poor. A good example of minimal stress occurs when you are about to fall asleep. As stress increases, so does the level of attention and motivation, resulting in improved performance. At very high levels of stress, however, panic ensues and performance deteriorates dramatically. This is particularly true when you have to perform complex or unfamiliar tasks that require a high level of attention. That is, the more difficult the task, the more likely it is to be affected by stress.

TYPES OF STRESS

There are two broad categories of stress: chronic and acute. **Chronic** stress is the result of long-term demands placed on the body by both positive and negative major life events. Examples of these events are marriage, the death of a family member, divorce, or prolonged concern over job security or health.

Acute stress results from demands placed on the body by current issues. Examples from flying include encountering an unexpected windshear on landing, or a higher-than-expected

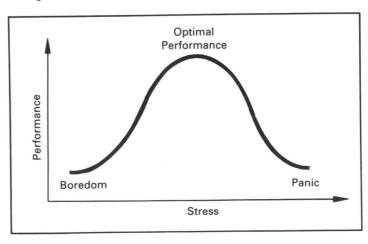

Figure 7-1

headwind forcing you to consider an alternate destination. Non flight-related acute stress may occur if you lose your wallet, have an automobile accident, or cut your finger.

Both types of stress have an impact on your ability to fly. Very often, the presence of chronic stress can make a situation that you can normally deal with more difficult to handle. That is, chronic stress can exaggerate the effects of acute stress. This means that, if you want to learn about the effects of stress on your flying, you have to look at all the stresses in your life and not just those directly related to aviation.

EFFECTS OF STRESS

We alluded earlier in the chapter to the fact that stress has a greater impact on more difficult tasks. Except in extreme cases, stress has little impact on your ability to physically manipulate the flight controls. It can, however, have a dramatic impact on complex tasks, such as scanning the instruments or making logical decisions. A common effect of excessive stress is fixation or tunnel vision, where you focus on one problem to the exclusion of others. You lose your ability to see all of the information in front of you, making it difficult or impossible to make sound choices from the available alternatives.

As your workload increases, stress makes it increasingly difficult for you to handle all the demands. Figure 7-2 is a "Margin of Safety" diagram that shows the difference between

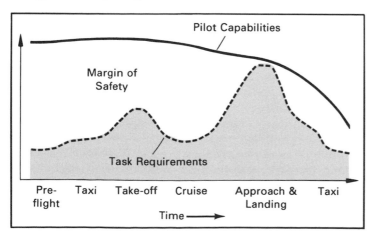

Figure 7-2

pilot capabilities and task requirements. The margin of safety is minimal during an approach when task requirements are highest. This is one of the reasons why accidents often occur in this phase of flight. The situation is often made worse because the end of a flight is where your ability to deal with problems is at its lowest due to physical and mental fatigue.

One other significant and universal effect of stress is a reduction in verbal communication. As stress increases, you are likely to withdraw more and more into yourself, isolating yourself from other people. This can be a significant problem in aviation, because good communication can make the difference between a safe or unsafe flight. The impact of poor communication becomes even more critical in a multi-person crew environment.

STAGES OF STRESS

Your body responds to demands made upon it in three stages. First, there is an alarm reaction; then resistance; and, finally, exhaustion (if the demand continues). This three-stage response is part of the human's primitive, biological coping mechanism which may have prepared our ancestors for the "fight or flight response."

In the **alarm stage**, your body recognizes the stressor and prepares to deal with it through either confrontation or fleeing. Your brain stimulates the pituitary gland to release hormones. These hormones trigger the adrenal glands to secrete adrenaline into the bloodstream. Adrenaline increases your heartbeat, rate of breathing, and perspiration. It raises your blood sugar level, dilates the pupils, and slows digestion. You also may experience a huge burst of energy, greater muscular strength, as well as improved hearing, vision, and alertness. All of this leads to an increased ability to find a solution to the problem causing the alarm.

If fear is the predominant emotion, the body reacts by lowering the blood pressure, resulting in a pale face. You may recall such an alarm reaction during your beginning flights when, for example, you felt a sudden buffeting close to the runway on landing. You may also recall the effects of your body's alarm reaction, such as an increased heartbeat, sweaty palms, breathlessness, and perhaps even light-headedness from the lowered blood pressure.

If anger is the predominant emotion of the alarm stage, the body secretes nor-adrenaline, resulting in high blood pressure

These people are characterized by a competitive, aggressive, achievement-oriented, time-dominated orientation to life. Since this behavior is condoned and applauded by our achievement-oriented society, Type-A people are usually not aware that their behavior is detrimental to their health and well-being. They are also not aware that their behavior often creates problems for other people, who want to take things more slowly.

The behavior of a **Type-B** person is everything Type-A people reject. Type-B individuals have found a comfortable, more relaxed pace. They look at scenery with enjoyment, allow time for frequent refreshment and rest stops, and really enjoy being alone or with friends and family. Type-B people work more slowly and thoughtfully, often resulting in greater creativity. They allow themselves the leisure to develop more fully as people and have a number of outside interests, activities, and friendships. Many Type-B personalities have plenty of drive and achievements, but time is scheduled with a calendar, not a stopwatch. If you recognize the Type-A pattern in yourself, you should consider modifying your life style. It will quite likely contribute to a longer and healthier life.

It is tempting to say that one personality type is more likely to result in a safer pilot, but we are not aware of any conclusive data to support such a statement.

HOW MUCH STRESS IS IN YOUR LIFE?

If you hope to succeed at reducing the stress associated with flying or other parts of your life, it is essential you begin by making a personal assessment of stress in all areas of your life. You may face major stressors, such as the death of a family member or a change in residence, or the birth of a baby, as well as a multitude of comparatively minor stressors like a social event or a vacation. These major and minor stressors have a cumulative effect that constitutes your total stress-adaptation level, which can vary from year to year.

THE LIFE EVENTS STRESS PROFILE

To enhance your awareness of the level and sources of stress in your life, complete the Life Events Stress Profile questionnaire in figure 7-3. Circle each event you have experienced and a red face. Anger is a dangerous emotion in flight because the high blood pressure it triggers prevents you from thinking clearly and, thereby, lowers your ability to develop a solution to the problem causing the alarm.

The long-term effects of a fear-induced alarm stage are not harmful unless it is very severe and lasting. On the other hand, the long-term effects of an anger-produced alarm stage can be dangerous because of the problems created by high blood pressure.

In the **resistance stage**, your body repairs any physical or mental damage caused by the stressor. In some cases, the body adapts to stresses, such as extreme cold, hard physical labor, or worries. Fortunately, most physical and emotional stressors last for only a short duration, while your body simply copes with physiological stress. During your lifetime, you will go through the first two stages many times, and they are needed to help you deal with the many demands and threats of daily living.

If the stressor continues, your body will remain in the alarm stage in a constant state of readiness for a prolonged time. For example, if you are caught above clouds flying VFR or realize that you may not reach your destination because of a fuel shortage, your alarm may continue for several hours. Eventually, your body may be unable to keep up with the demands, leading to the final stage of stress: **exhaustion**. With exhaustion, almost all control is lost as your mind no longer is able to keep a proper perspective. Sometimes, exhaustion can become so extreme that you may give up trying to solve the problem and resign yourself to your fate. This can be a very dangerous situation.

THE EFFECTS OF PERSONALITY

There is no question that your personality influences the way that you react to stress. Some people have personality types that contribute to stress-related disorders. They may feel so fearful of making mistakes, being criticized, or doing less than a perfect job that they withdraw from challenging situations and avoid confrontations. This usually leaves them feeling unfulfilled, frustrated, and incompetent.

Cardiologists describe two personality types that are linked with certain diseases. **Type-A** people have behavior patterns that are seen as a major cause of coronary heart disease.

in the last 12 months, and sum the Life Change Units (LCU's) associated with each.

Life Change Units	Life Event
LIFE EVENTS STRESS PROFILE	
100	Death of spouse
73	Divorce
65	Marital separation
63	Jail term
63	Death of close family member
53	Personal injury or illness
50	Marriage
47	Lost your job
45	Marital reconciliation
45	Retirement
44	Change in health of family member
40	Pregnancy
39	Sex difficulties
39	Gain of new family member
39	Business difficulties
38	Change in financial state
37	Death of close friend
36	Change to different line of work
35	Change in number of arguments with spouse or partner
31	Mortgage or loan over $30,000
30	Foreclosure of mortgage or loan
29	Change in responsibilities at work
29	Son or daughter leaving home
29	Trouble with in-laws or partner's family
28	Outstanding personal achievement
26	Spouse or partner begins or stops work
26	You begin or end work
25	Change in living conditions
24	Revision of personal habits
23	Trouble with boss or instructor
20	Change in work hours or conditions
20	Change in residence
20	Change in school or teaching institution
19	Change in recreational activities
19	Change in church activities
18	Change in social activities
17	Mortgage or loan less than $30,000
16	Change in sleeping habits
15	Change in number of family social events
15	Change in eating habits
13	Vacation
12	Christmas
11	Minor violations of the law
Total Life Change Units _____	

Figure 7-3

The more change you have in your life, the more stress you will experience and the more likely you will suffer a decline in health. In a pilot study of Navy personnel, it was found that of those people who reported LCU's that totaled between 150 and 199, 37% had associated health changes within two years. Of those who had between 200 and 299 LCU's, 51% reported health changes; and for those with over 300 LCU's, 79% had associated injuries or illnesses to report within a year of the life crisis. It was also found that there was a very high correlation between a high LCU score and the accident rate among pilots.

Although these data are very revealing, it does not necessarily mean you will get sick or have an accident if you have a high LCU score. Each of us has a different stress adaptation capability. That is, stress that incapacitates one person may have no effect on another. When you exceed your personal adaptation level, stress overload may lead to poor health, illness, or accidents. To avoid exceeding your personal limit, learn to recognize the warning signals from your body and take heed of those that tell you when your stress level is getting too high. When you observe warning signs, do not procrastinate; take preventative action immediately.

TIME AND STRESS

The urgency of time drives most of us. When piloting an aircraft, these time pressures can be very apparent. In flight, there are often times when you have to perform multiple tasks simultaneously. The amount of fuel remaining is directly related to time. Flight schedules, passenger requirements, and economy of operation are usually related to time. In flying, demands often exceed the time available, and overloading means your stress response is likely to be dangerously high.

Most people have a fairly well-defined sense of time urgency within which they work effectively and gain a sense of accomplishment. If you go beyond this comfort zone of reasonable time pressure, you are likely to feel threatened by deadlines, and time seems to run out. Furthermore, it is difficult to set aside enough recovery time for a change of pace. As a result, you begin to feel over-stressed and become irritable. You are also likely to suffer impaired judgment, hypertension, headaches, and indigestion, which are frequent early signs of distress and potential illness.

RECOGNIZING YOUR WARNING SIGNS

As we mentioned above, a common source of stress is always racing against time — a condition called **chronic overload**. To minimize this as a source of stress, you must first learn to recognize your personal warning signs, then take timely action to reduce its effects. The following exercise can help reveal whether you are suffering from the symptoms of chronic overload. Answer each of the questions. Do you:

1.   rush your speech?

2.   hurry or complete other people's sentences?

3.   hurry when you eat?

4.   hate to wait in line?

5.   never seem to catch up?

6.   schedule more activities than you have time available?

7.   detest wasting time?

8.   drive too fast most of the time?

9.   often try to do several things at once?

10.   become impatient if others are too slow?

11.   have little time for relaxation, intimacy, or enjoying your environment?

If you answered "yes" to most of the questions, you may well be in a phase where you are suffering from chronic overload. This does not necessarily mean you have a problem, because people handle time stress differently. Most people go back and forth between being stressed by time and having a more relaxed schedule. Some people can and do live fast lives, because their bodies and minds can handle a faster pace. Others learn to adjust to a faster pace. However, the chances of stress are greater in the fast lane, especially if you are not aware of its dangers or do little or nothing about the warning signs.

Common ways of handling time pressure are to juggle several activities at once or to attempt to cram several activities into insufficient time. Unfortunately, these are rarely good strategies because the human brain is poor at performing many simultaneous conscious operations efficiently, largely because the requirements of one task interfere with those of the others. When you have too much to do, you start losing track of important information and you increase the chances of making mistakes.

Accident data suggest that most mishaps result not from a single catastrophic decision, but from a series of poor decisions, none of which may be critical by itself. This is called **The Poor Judgment (PJ) Chain**, where one bad decision increases the probability of another. As the PJ chain grows, the chances for a safe conclusion to the flight decrease. Research also shows that each error leads to an increase of workload (to rectify the error). When this increased workload is compounded with the knowledge that you have made a mistake, your stress level rises greatly which, in turn, makes you more prone to making errors. The best strategy for breaking out of this vicious circle is to acknowledge that you have made an error and take action immediately to correct it. Procrastination will only lead to more errors, making the situation worse.

LEAD-TIME AND AFTER-BURN

Associated with any activity are two necessary time periods called lead-time and after-burn. Consider, for example, the situation in which you have a flight test coming up. **Lead-time**, or the time before the test, may result in anticipatory stress which is often beneficial in moderate amounts. The stress often helps you prepare both your body and mind for what is about to happen. It increases sharpness and motivation. However, it can be an interference if you pay more attention to what may happen during and after the test than you do to the preparation. When this reaction is excessive, it can cause you to fall behind a reasonable study schedule.

In a flight situation, if you use the lead-time to worry about what will happen if you do not solve a particular problem, you are likely to "get behind" the demands of your flight. This, in turn, can cause more time pressure and distress.

After-burn time, on the other hand, is the time needed after the test to think about results and to put the experience to rest. If there is not enough time to "come down" — to relieve the tensions built up during the lead-time and the review — the energy that surged during the experience will not be released, and your body and mind will remain stressed. A fast pace can be a significant source of tension, stress, and disease, especially if it is led by someone who needs quite a bit of lead-time and after-burn time. You may well have experienced this in the form of a sore back or stiff neck after a stressful event.

COPING WITH STRESS

Earlier in the chapter we categorized stress as acute or chronic. Acute stress results from what is happening at the moment, while chronic stress is the product of an entire lifestyle. Just as the causes of each type are different, so are the coping strategies.

ACUTE STRESS

Of the two types, acute stress is the easiest to counter. If you feel stressed, you can often remedy the situation by simply taking a five-minute break and relaxing. This helps reduce the immediate overload and allows you to develop a better perspective on whatever is causing the stress. There are times in the cockpit that this is not possible, such as when you are icing up. In this case, slow, deep breathing can be helpful.

You also can use this technique to prevent the onset of acute stress. If you expect a situation where the workload and associated stress will be high, plan to take a mental break before it occurs. For example, before you arrive at an unfamiliar airport, review the airport information, runway orientations, pattern directions, and radio frequencies. Then, take a few minutes to relax while your workload is low and prepare mentally to deal with any problems that may arise. If you are relaxed and mentally prepared, an unexpected change, such as in the active runway, will not create excessive stress.

CHRONIC STRESS

Chronic stress is not just the product of an occasional crisis but, rather the result of long-term issues. Its treatment is usually long-term as well, requiring discipline and dedication.

One of the best ways we have found to cope with chronic stress is to use a concept called the "total body approach" or the "wellness concept." The objective of this concept is to attack the problem before it becomes serious. The total body approach takes into account all six aspects of well-being:

1.   Physiological
2.   Nutritional
3.   Environmental
4.   Emotional
5.   Spiritual
6.   Lifestyle values

These areas are interrelated. What you do in one affects the others. This is true for both creating stress and coping with it. For instance, poor eating habits can increase your stress level, leading to weight problems and lack of vitality. This lack of energy can slow down productivity at home and work and lead to increased pressure to get things done. The pressure could lower your self-esteem or create defensive behaviors which can throw your entire lifestyle out of balance. As your stress increases to unhealthy levels, you may be more inclined to eat poorly, and the cycle repeats itself.

Similarly, once you take a positive approach to coping with stress, successes in any one area have a positive effect on the others, making your efforts seem more worthwhile. The difficult part of dealing with chronic stress is summoning the will to make a start. That decision, itself, can increase the stress you are feeling.

Before we discuss the actions you can take for each of the six aspects of well-being, we want to make it clear there is no single, magical solution. The selection of specific ways of managing stress is a matter of individual choice and circumstances. What works for a friend may not work for you. It is important, therefore, to consider how different courses of action are suited for you.

As you consider different courses of action, asking yourself the following two questions can help bring a better perspective to your deliberations.

1.   In what ways will a particular choice or action promote my own good health and minimize stress?

2.   In what ways will my efforts promote the health and development of others and reduce stress?

By considering change in this positive light, you actually can mobilize stress as a positive force.

During the process of seeking relief from your stressors, you should try to avoid "addictive" solutions, because they often generate their own set of problems. Some of the potentially destructive solutions include violence, procrastination, drug abuse, overwork, compulsive spending and gambling, total withdrawal, and caustic remarks. These approaches are likely to make your problems worse or, perhaps, create new ones.

Living with your stress is perhaps the least acceptable of all the possible solutions. This may be a necessary action for short periods of time, but it does not promote long-term health. For instance, the intensity with which you prepare for a flight check is both invigorating and grueling. A temporary sacrifice is made to reach a goal and does little lasting harm. Unfortunately, some people are almost addicted to stress and go to great lengths to create stressful situations. These people not only live with stress, they wallow in it, playing out either "loser" and "poor me" lifescripts or trying to show how tough they are.

PHYSIOLOGICAL ACTION

One of the easiest actions you can take to reduce the effects of stress is to increase health-producing activities, such as relaxation and exercise. The relaxation can bring you increased peace of mind and result in a better perspective on your stressors. The exercise can make you feel healthier, resulting in a better self image and, consequently, a better outlook on life.

Relaxation

The following relaxation techniques are useful for everyone, but particularly for those who suffer physical symptoms such as headaches, backaches, stiff necks, tense or rigid bodies, ulcers, or high blood pressure. These techniques take some time to learn, and the effectiveness of each is dependent upon regular use.

You may feel self-conscious when you first use some of these techniques, because they involve talking or thinking to yourself. For this reason, you may want to do the exercises in private. When you become used to them, the self-consciousness is likely to disappear.

The first exercise is called deep muscle relaxation. It is a passive process that involves getting yourself into a relaxed position on a comfortable chair or sofa, and then focusing your attention on various muscle groups throughout the body. First, you tense, then release each group of muscles while saying to yourself, "Relax, relax, relax." This builds up an association between the mental process and physical relaxation. Eventually, this technique will enable you to automatically relax whenever you think, "Relax."

You can use a similar technique in the cockpit but must always take care that it does not distract you from your primary task of flying safely. In the cockpit, it is helpful just to tense and relax your muscles, even though your attention is focused on flying and not creating a meditative state of mind.

Progressive relaxation is a similar technique, except you do not tense your muscles. Instead, you mentally suggest relaxation through thoughts like, "My feet are completely relaxed; my feet are completely relaxed," while consciously relaxing your foot muscles. Progressive relaxation is often accompanied by deep breathing or visualization techniques.

Deep breathing exercises are powerful techniques for reducing tension and can be done anywhere. When done for an extended time, such as ten minutes or more, they can produce a deep state of calmness and relaxation, making it difficult for your emotions to become aroused out of a tranquil state. If you are flying, shorter sessions of deep breathing will help to relax you without affecting your concentration. Several disciplines include breathing exercises as part of their relaxation strategies. In yoga, "pranayama," or control of the life force, is an important study. Since we all have to breathe anyway, controlling your breathing is a quick and easy way to help relieve tension.

Exercise

Exercise reduces physical tension and anxiety and increases the quality of our lives. Flying, for the most part, is a sedentary activity. We sit in an aircraft for long periods without any physical exertion. Health researchers strongly recommend a weekly exercise routine of at least three, 30-minute periods of some vigorous workout. However, they add two important cautions:

1. Strenuous exertion by anyone who is normally sedentary or overweight can be hazardous. For such persons, a medical evaluation is essential before beginning an exercise program.

2. Exercise, alone, will not reduce risk of coronary heart disease.

NUTRITIONAL ACTION

The overall strategy in this area is minimizing or stopping activities that are detrimental to your health, such as drinking or eating unhealthy food. We discussed the effects of diet on the body in the previous chapter and need not repeat the information here.

ENVIRONMENTAL ACTION

Environmental action refers to taking steps to control the places in which you live and work. The most common of these actions is withdrawal, which is designed to remove you from the stressful situation. This is often used when other approaches have not been successful. Leaving a stressful situation for a while by taking a walk, a day off, a nap, or a vacation can be healthy responses to restore vitality and relieve overload.

Like other coping responses, withdrawal can also be destructive, depending upon when and how it is used. It can provide a change of pace and renewal, or it may merely be a means of escape and, in fact, create more stress. When withdrawal is used to escape, it is often called avoidance, because it means you are not really addressing the problem.

Avoidance techniques rarely solve any problem in the long run. You can use them for temporary relief from a situation but, if carried on for too long, they can create further problems.

Extreme examples of withdrawal, such as changing your job, moving out of town, or ending a relationship should be considered as a last resort in most cases. It is only when a major stressor cannot be changed (for instance, the crowding and noise of city life) and no other solution found, that total withdrawal may be the most viable coping response. Usually, there are alternatives by which you can moderate the negative stressors in your life without leaving the stressful situation completely.

Although some situations are beyond your control (such as the erratic movement of a low pressure system into your flight path or an unsigned maintenance release for your aircraft), this is not true for all stressors. You can often change features in your environment that cause you stress. This is particularly true if the stress you are feeling is caused by other people, such as colleagues at work or a spouse. In this case, one of the best environmental actions is to make some changes in your own behavior, rather than trying to change everyone else. The following behavioral changes are usually beneficial.

1.   Develop new skills.
2.   Establish a support network of close friends.
3.   Learn to be more diplomatic.

4. Learn to be more assertive.

5. Be more tolerant of others' imperfections and of your own.

6. Broaden your perspective of the stressful environment.

If you take this approach and work on changing your behavior, what you are doing is taking control over the relationship between you and the demands that are causing the stress. For example, the source of your tension may be an over-demanding chief pilot who assigns you a flight late in the day when you are already fatigued, and then berates you for falling behind schedule. Being more assertive and candid with your boss about your work schedule and time demands, suggesting alternatives, and understanding the chief pilot's scheduling problems could defuse the tension.

As you seek to change your environment, it is wise to remember to try to change only those things you can and to accept those you cannot by tolerating them and recognizing your own limitations.

EMOTIONAL ACTIONS

Emotional action usually means working on your attitudes and outlook on life. Loosening up inhibitions, overcoming your limitations, and working on developing a positive attitude toward life are all important aspects of a well-rounded, stress-controlled existence. These strategies are helpful in dealing with daily stress levels, but they also reduce the chance of high stress levels and panic in flight situations.

One of the best changes is to develop a positive attitude toward life. Rather than worrying about everything that is causing you stress, try putting the stressors in a favorable context. Remember, some stress is useful or necessary. Recognize the beneficial aspects of stress, and use the power of positive thinking to turn the stress to your advantage. Your attitude determines whether you perceive any experience as pleasant or unpleasant. For example, if you find yourself always complaining about how well your airplane is cleaned, stop complaining and start working with the maintenance people on establishing appropriate standards and procedures. You may even want to do some of the cleaning yourself for a while to establish a good role model.

In some extreme cases of stress, you may want to seek professional counseling and therapy. This can help you develop a better perspective on what is troubling you and may result in the development of good coping strategies.

SPIRITUAL ACTIONS

Becoming more involved in your religious faith and practice can be a good coping strategy, because it provides both a support system and unconditional acceptance. Very often the community in which you worship provides a new set of friends who judge you by what you are to them and not what you are in your home or work life. This can free you from many of the daily tensions and allow you to have stress-free time that is like a mini-vacation. Moreover, the very nature of worship can help you reestablish your own perspective of yourself, which is helpful in making the decisions necessary to deal with your stress.

LIFESTYLE ACTIONS

In a sense, if you take actions on the other aspects of your life, you have already taken action on your lifestyle. From the perspective of reducing stress, one of the most powerful lifestyle actions is to slow down. Do things in moderation, and spend more time relaxing with your family and friends. Give yourself more time for the things you like.

PREFLIGHTING YOURSELF

Most pilots give their aircraft a thorough preflight inspection to ensure all systems are working. Yet, many pilots forget to preflight themselves. The **I'M SAFE** checklist in figure 7-4 is a short summary of the material presented in this chapter. It may be useful to you as a part of your flight preparation. If you answer "Yes" to any of the questions, you should reevaluate your decision to fly.

CONCLUSION

As a pilot, you are subject to three types of stress, physiological, physical, and emotional. Each of them can adversely affect your performance if present in excess. Emotional stress is the most difficult to deal with, because it is usually caused by situations not associated with your flying. It is, therefore, easy to ignore and can often become chronic.

ARE YOU FIT TO FLY? THE "I'M SAFE" CHECKLIST	
Illness?	Do I have any symptoms?
Medication?	Have I been taking prescription or over-the-counter drugs?
Stress?	Am I under psychological pressure from the job? Do I have money, health or family problems?
Alcohol?	Have I had anything to drink in the last 24 hours? Do I have a hang-over?
Fatigue?	How much time since my last flight? Did I sleep well last night and am I adequately rested?
Eating?	Have I eaten enough of the proper foods to keep me adequately nourished during the entire flight?

Figure 7-4

The effects of stress are cumulative, so any stressful experience in the airplane is made worse if chronic stress is present. As stress rises, it is easier to become overloaded, and you will be less able to deal with all the demands being placed upon you. In situations of high stress, you are more likely to make poor decisions, such as pressing on into bad weather or overflying good landing areas until you are out of fuel.

Stress can grow to the point where the burden is intolerable unless you know how to cope. Although there is no panacea for managing stress, your chances of success are increased if you take an overall "total-body" approach. Taking action on all areas of your life has better results than focusing on just one or two.

Stress affects everyone. It is far better to know how to make it an ally rather than letting it be your enemy.

Exercises

Chapter Questions

1. Why are emotions important to safe flight?

2. Why is the understanding of stress important to pilots?

3. Name and describe the two types of stress.

4.   Name and describe the three stages of the stress.

5.   Many different stress coping responses are given in the chapter. List a few that you can use.

6.   List and describe the three sources of stress from the flying task.

REFERENCES AND RECOMMENDED READINGS

Benson, H. 1975. *The Relaxation Response.* New York: William Morrow.

Diehl, A.E., Hwoschinsky, P.V., Lawton, R.S., and Livack, G.S. (Eds). 1987. *Aeronautical Decision Making for Student and Private Pilots.* DOT/FAA/PM-86/41. Washington, DC: Federal Aviation Administration.

Holmes, T. & R.H. Rahe. 1967. The social readjustment rating scale. *Journal of Psychosomatic Research*, 11, 213-218.

McQuade, W. & A. Aikman. 1975. *Stress.* New York: Bantam.

Sellye, H. 1974. *Stress without distress.* Philadelphia, PA: Lippincott.

Pilot Judgment 8

INTRODUCTION

In the previous five chapters, we discussed how pilots obtain and process information and how their performance is affected by factors such as cockpit design, and mental and physical health. In this chapter, we deal with **judgment**, which we regard as the process that ties all the previous parts together.

All too frequently we read of flying accidents caused by bad pilot judgment. Sometimes pilots run out of fuel; sometimes they try to take off with more load than the plane is designed to carry; more often, they fly into known adverse weather conditions. In fact, of the pilot-caused accidents in civil aviation, over half are the result of bad pilot judgment. From such reports, it is easy to believe that some pilots have innately poor judgment. We do not agree with this impression; we believe that judgment is a skill that can be learned.

WHAT IS JUDGMENT?

Before we define judgment, we will set the stage by giving a few examples that illustrate what it is. A cartoon by Brickman called "The Small Society" once showed two passengers walking down an airport terminal corridor. One says to the other, "My plane came in late because the pilot didn't like the sound of the engine during warm-up. We had to taxi back to change pilots."

Although this is a humorous story that does not seem plausible in the real world, in fact pilots do have differing levels of judgment. Our personalities and attitudes toward risk-taking, combined with pressures on us from others, may cause us as pilots to make decisions that would be considered stupid by others. The following true story illustrates:

> A pilot for a large university transportation service was asked to make a flight to a midwestern city. Drawing upon his considerable experience and good judgment, he determined that the weather was not sufficiently safe for such a trip in a single-engine aircraft because there was moderate icing reported in the clouds. His chief pilot, a personal friend of the passengers, who had persuaded them to fly instead of drive, was angry with the pilot's decision to cancel the flight, considering the passengers had already arrived at the airport. The chief pilot himself hastily loaded the passengers into the aircraft and took off on the original flight plan for the destination. Nearing the destination uneventfully, he was asked to hold for an approach clearance; at that time, an almost routine occurrence at this busy airport with adverse weather.

It was at this point that he realized he had not fueled the aircraft prior to takeoff and was quite uncertain about the amount of fuel remaining. Not wanting to appear concerned, however (He had the "Right Stuff."), and feeling that there is "always" more fuel left than the gauges show, he sought no special handling until the engine quit due to fuel starvation. Luckily, he managed to crash-land the airplane in a field near a gravel quarry. He came away unhurt but, tragically, one of his passengers was injured and disabled for life.

In contrast, consider the case of another pilot, who was flying in Zimbabwe in southern Africa:

The flight started in Kariba in the north of the country, went to a short, dirt strip in a game park called Mana Pools, followed by a flight to a large airport next to the Hwange Game Park. After Hwange, the final leg was a relatively short flight to Bulawayo. The pilot estimated the trip to Hwange would take about three hours and, consequently, ensured that there was three hours and forty-five minutes of fuel on board. He did not want to top off the tanks because of the high temperatures and the short strip at Mana Pools. He wanted to minimize his weight for this takeoff.

The first part of the trip went uneventfully, and he landed safely at Hwange Airport. He found to his surprise, however, that there was no fuel there. The nearest fuel was at Victoria Falls, approximately 30 minutes flying time away.

He decided to find a car to drive to Victoria Falls to buy enough fuel to let them fly safely for refueling. This caused a whole day's delay in the trip, resulting in some of his passengers losing consulting income and others missing important meetings.

The situations the two pilots faced were similar, but they acted very differently. In the first case, the pilot felt pressured by social obligations to fly in adverse conditions and, in doing so, neglected elementary safety checks. The other pilot reasoned that even if he left his passengers behind and flew himself to Victoria Falls there was an unacceptable level of risk. He decided to drive instead.

What made these two pilots act so differently? The contrast of these two stories is what our discussion of judgment is all about. In one case the pilot exhibited bad judgment and in the other good judgment.

So what is pilot judgment? We define it as follows:

> **Pilot judgment** is the mental process that we use in making decisions.

We have seen that pilots obtain most of their information through the eyes, ears, touch, and vestibular senses. Each of these sensory systems makes changes to the information coming in to reduce it to manageable form for the brain. The brain further reduces the information by placing it into known categories to make it easier to use. The brain then analyzes this information, comes to a decision, and initiates action. This whole process, from gathering information to taking action, is called pilot judgment.

PERCEPTUAL JUDGMENT

As we mentioned in the earlier chapters on the brain and the eyes, pilots are constantly making decisions based on their visual perceptions. For example, they judge distance, clearance, altitude, closure rate, and speed. These perceptual judgments are highly important to the pilot's task of controlling the aircraft. However, they do not require much thought and are relatively easy to learn and to perform consistently, particularly when the pilot is aware of the anomalies and illusions that can occur.

COGNITIVE JUDGMENT

In this chapter, we concentrate on more complex forms of judgment that we call cognitive judgment. In comparison with perceptual judgment, the characteristics of cognitive judgment are:

* The information available is more uncertain.
* The pilot has more time to think.
* There are usually more than two alternatives.
* The risks associated with each alternative are harder to assess.
* The final decision is more easily influenced by nonflight factors such as stress, fatigue, financial pressure, personal commitment, and so on.

The two forms of judgment can be seen as two ends of a continuum of cognitive complexity as shown below:

PERCEPTUAL JUDGMENT   COGNITIVE JUDGMENT

➤ **Increasing Cognitive Complexity** ➤

Many decisions that we make early in our flying careers, when we have little training or experience, are cognitive — that is, they require a considerable amount of thought. Later, with experience, these decisions become perceptual. For example, when you first learn to land the airplane, you have to think, "Am I too high, or too low?" Your instructor may tell you, "You're too high, lower the nose!" Those thoughts and instructions are cognitive.

It does not take much training time before you are able to look at the view and know, almost automatically and without thought, that you are too high. This behavior is perceptual. A characteristic of highly experienced professional pilots is that most of their decisions are automatic. It is one of the goals of this chapter to help you move from cognitive to perceptual judgment. That is, we want you to make good judgments automatically. For the remainder of the chapter, unless otherwise stated, when we say judgment we refer to cognitive judgment.

Our definition tells us that judgment is a process, meaning it requires a series of related steps to complete. Numerous models of this process have been offered. We will consider one that has several steps. Understanding each step will make you more aware of what you can do to develop better judgment. Our model comprises the eight steps shown below:

1.   Vigilance
2.   Problem Discovery
3.   Problem Diagnosis
4.   Alternative Generation
5.   Risk Analysis
6.   Background Problem
7.   Decision
8.   Action

Vigilance. The first element of the judgment process is vigilance. The pilot must look for and expect problems that may affect the safety of the flight at all times. For example, some flight instructors tell their multi-engine students that, to be prepared for such an emergency, they should expect to have an engine failure on every takeoff. In several of the earlier chapters, we discussed how expectations affect our ability to perform. This is an example of how we can turn what is often a liability into a benefit. This step requires observation skills.

A less obvious example of good pilot vigilance is maintaining a state of awareness of potential problems. For example, if weather is deteriorating ahead, the pilot must decide whether to continue the flight. Being aware of the negative consequences of such a decision may prevent an error in judgment. Vigilance for what may go wrong is just as important as noticing that something has gone wrong. In addition, awareness of potential problems allows the pilot to plan for alternatives ahead of time during low levels of stress, making the correct decision more likely — and bringing about lower levels of overall stress. More will be said about this as we discuss the use of the DECIDE technique in pilot decision-making later in the chapter.

Problem discovery. The second step of the judgment process is problem discovery. At this stage, a change in the status of the flight is discovered that could affect its safety. This step requires curiosity as well as perceptual skills.

Problem diagnosis. The third step of the judgment process is problem diagnosis. During this stage, the pilot tries to discover the nature of the problem. This requires information processing, knowledge, memory, and problem-solving skills.

Alternative generation. The fourth step of the judgment process is alternative generation. During this stage, the pilot identifies a series of alternatives that may be available to solve or to avoid the problem. This step requires creativity and knowledge.

Risk analysis. The fifth step of the judgment process is risk analysis. During this stage, the pilot determines the risks involved in taking each of the identified alternatives. This step requires computational skills.

Background problem. The sixth step of the judgment process is called the background problem. This step is more than just a stage in the process. It is like a cloud that covers the whole process, but is particularly important just as a decision is about to be made. It refers to the nonsafety related factors that affect every cognitive decision made by the pilot, such as economics, commitment, ego, peer pressure, illness, and fatigue. To handle these pressures requires a considerable amount of mental fortitude and discipline.

Decision. The seventh step of the judgment process is the decision. At this stage, the pilot's mind is made up. The pilot is now ready to act. This step requires leadership skills.

Action. The final step of the judgment process is action. During this stage the pilot acts on the decision, which may involve actions such as moving the controls or telling the passengers that they will not be able to fly today. This step requires eye-hand coordination and skills dealing with people.

So judgment is a process comprising a number of steps. The eight steps of the judgment model described above help us to see the basic human abilities that are needed for good judgment. These include observation, perception, curiosity, information processing, knowledge, problem solving, creativity, computation, mental fortitude, discipline, leadership, and social skills. Being aware of the steps in making judgments and knowing the abilities needed for them will make you a better pilot.

WHY IS JUDGMENT SO IMPORTANT?

Judgment is important in flying because regulatory bodies, such as the FAA, have written the regulations in such a way that pilots are expected to exercise judgment in all phases of flight. In the regulations, you as a pilot are given an enormous amount of responsibility for your safety and that of others. This responsibility manifests itself in the following ways:

1. **Command of the Aircraft**: Pilots are expected to be in command of their aircraft at all times. This is clearly stated in the regulations. This means, for example, that you should not give up your authority to make decisions to demanding passengers by letting them unduly influence your judgment.

2. **Adherence to the Regulations**: Pilots are expected to adhere to the regulations. This is so basic that it should not need to be mentioned. But, flying by yourself, you are not being monitored by anyone, and it is easy to think you can bypass or stretch the rules. Remember, other pilots fly by the rules and expect you to do the same. For example, in VFR conditions you are expected to fly at certain altitudes and visually check for traffic. Overtaken traffic expects overtaking traffic to follow this rule, in particular.

3. **Interpretation of the Regulations**: The FAA regulations are offered under the expectation that pilots will interpret them based on their own skill and aircraft capabilities. Because the FAA has the dual purpose of promoting aviation and regulating it for safety, it aims its restrictive regulations at the operator with the most skill and equipment capability — not the one with the least. The FAA regulations represent minimum standards for safe flight which means that to be safe, all pilots, except perhaps the most proficient, must be more conservative in decision making than the established standards. In the USA, the FAA testing program (for the certification of aircraft equipment and flight procedures) uses the best pilots and equipment to test all operations that it regulates, not mediocre pilots and

equipment. Although most pilots rarely consider themselves to be anything but good, few have the skill of these test pilots. Therefore, the decision to attempt a certain maneuver (for example, a minimums approach), is left to you, the pilot. You alone are responsible for assessing your own skill level, state of fatigue and health, and the reliability of your equipment.

You should establish personal minimums for all types of flying. Just as the airlines do, when you are new to an aircraft, you should set your IFR approach minimums higher than those established by the FAA, perhaps add 200 feet. Your minimums for crosswind possibly should be set higher than those shown in the aircraft flight manual. In planning your flights, be conservative on fuel estimates and runway length requirements and remember that the values found in the flight manual were established by test pilots who had the best information and equipment possible. You are most likely lacking in both. As you gain experience with your airplane, you can move your minimums closer to those set by the FAA and the flight manual.

In all phases of flight, therefore, a great deal of the decision making is up to you. The whole system is based on the premise that pilots will exercise good judgment in looking out for their own safety as well as for the safety of other aircraft and people. That is, it is a system based on trust — the trust that you will respect the responsibility that has been given to you and will not flaunt it just because there is no one to monitor what you are doing.

GOOD JUDGMENT CAN BE LEARNED

The traditional approach to flight training is to teach student pilots the capabilities and flight characteristics of an aircraft and its systems, knowledge of the national airspace system, general knowledge of meteorology, regulations, emergency procedures, and "stick and rudder" skills. The premise is that, if pilots have this kind of information, they will be able to exercise the good judgment required to assure safe flight.

Yet, as we have seen in our discussion of cockpit design, the senses, and the brain, there are many places where errors can occur in flight. Accident investigations reveal that many of these errors have unfortunate conclusions — they are listed as "pilot error" accidents, often in an attempt to attribute blame. In many cases, however, we do not regard such errors to be a result of negligence because there was no lack of ordinary vigilance and certainly no intent to err. The errors occurred only because of defects in the human information processing system, as discussed earlier in our chapter on the brain.

On the other hand, we should not just throw up our hands and exclaim, "Whatever will be, will be." We do have control of our destiny in most flying situations. With proper training, we can learn to recognize the defects in our information processing system, to take better control over our thought processes, and to improve our flying judgment. This is called learning.

Judgment is learned primarily from experience. It is not innate. All of our experiences in life mold the judgment tendencies we bring to our flight situations. These experiences can be useful or detrimental to good judgment in flight. On the other hand, most of our flying experiences develop good judgment — although there are some exceptions. In the rest of this chapter we show you how you can use your experiences and the experiences of others to develop good flying judgment.

There is an element of truth to the saying that good judgment is learned by the cautious and the lucky over a lifetime of flying experiences. However, neither caution nor luck are the best sources of judgment training for pilots. If you are too cautious, you may never know the limits of your capabilities. Obviously, we are suggesting that you test the limits of your capabilities only with a great deal of caution and preferably with a qualified instructor (i.e. aerobatic training). On the other hand, if you rely on luck, you may develop a belief that you are immortal and invulnerable to some of the real hazards of flight.

In reality, the best way to learn judgment is to discover it from the experiences of others. Although experience is a great teacher, in aviation the experience of others is safer. It is important to realize that good decision-making skills, like any other skill required in flying airplanes, can be learned more quickly and more safely through a systematic training program than through the traditional "trial-and-error" method.

We believe, therefore, that a more systematic and deliberate approach is needed — that judgment should be taught as a specific and essential skill. Our approach is to place students in situations that require them to grapple with a variety of circumstances requiring difficult decisions. As a pilot considers an action, the consequences of taking, as well as not taking, that action must be carefully considered. Judgment is learned through experience with as many decisions as possible. The place to gain that experience is in the classroom, computer terminal, or flight simulator where the only risk is pride — not in the air where there is risk of physical injury.

THE TWO-PART JUDGMENT MODEL

To help you understand how you can improve your judgment, it is helpful to reduce the eight-step judgment model into two parts. The first part, called **headwork**, comprises the first five steps of our judgment model, which relate to intellectual activities. The second part, called **attitude**, comprises the final three steps, which relate to motivation. In short, good pilot judgment is:

Part I: Headwork The ability to discover and establish the relevance of all available information relating to problems of flight, to diagnose these problems quickly, and to specify alternative courses of action and assess the risk associated with each.

Part II: Attitude The motivation to choose and execute a suitable course of action within the available time frame, where:
    a.   The choice could be either action or no action, and
    b.   "Suitable" is a choice consistent with societal norms.

HEADWORK

Headwork refers to the perceptual and intellectual ability of the pilot to detect, recognize, and diagnose problems, to determine available alternatives, and to determine the risk associated with each alternative. This part is purely rational, and if used alone, would allow problem solving in much the same manner as a computer. To reduce errors, headwork should be structured and orderly.

An excellent technique to improve your headwork is called DECIDE (Benner, 1975). The DECIDE technique has six steps that, when followed, can help to organize your thoughts and prevent overlooking facts that may be important. When faced with any decision involving uncertainty, you should remember the acronym DECIDE and do the following:

D — Detect: Detect the fact that a **change** has occurred that requires attention.

E — Estimate: Estimate the **significance** of the change to the flight.

C — Choose: Choose a **safe outcome** for the flight.

I — Identify: Identify **plausible actions and their risks** to control the change.

D — Do: **Do** the best option.

E — Evaluate: Evaluate the **effect of the action** on the change and on progress of the flight.

To use the DECIDE technique, first memorize the meaning of each of the terms. Then, each time a "change" (meaning any factor that is different from being ideal for flight) is detected, go through all of the DECIDE steps, considering each one carefully and doing what each suggests. As soon as you complete all of the steps, go back to the beginning and start again.

In practice, your initial thinking should be that your role is to be a vigilant monitor of all factors that could produce change during the flight. When such a change occurs, the judgment process is put into action by following each of the DECIDE steps. In practicing this technique, begin with decisions that have some element of uncertainty (such as, weather forecasts, fuel remaining, engine or navigation system reliability). As you repeatedly think through the model in these circumstances, it will become second nature to you and will help you in all your flight decisions.

Case Study

The following case study concerns the Air Illinois Flight 710 accident that occurred near Carbondale, Illinois, in 1983 and illustrates the use of the DECIDE technique. The accident involved an HS-748-2A that had a generator failure at night. Improper procedures were followed causing the disconnection of the second generator from the DC bus. These procedural errors were followed by a chain of poor crew decisions resulting in an attempt to reach the destination on battery power only. This attempt failed and the aircraft crashed several miles short of the destination when all DC electrical power was lost. A transcript of the cockpit voice recording of this flight is provided at the end of this chapter.

Figure 8-1 illustrates the use of the DECIDE technique in this accident. The Y's indicate that the item was followed and the N's indicate the item was not followed. In many cases, we must infer the thought process from the actions taken to assign a "Y" or an "N." The N's often suggest thought processes that are clues to the reasons for the accident.

Consider, for example, the third line of the script where the CHANGE is "Crew gets departure control offer to return to Springfield." The first letter is D (Detect). The crew clearly detected the change or offer because they quickly responded, rejecting the offer. The second letter is E (Estimate). We do not really know whether the crew estimated the significance of the offer, but we will give them the benefit of the doubt as trained airline pilots and indicate Y. The third letter is C (Choose). We know from the action (offer rejected) that they did not choose a safe outcome for this change; therefore, N. The fourth letter is I (Identify). Whether or not they considered other options, they certainly did not identify the option that would lead to a safe outcome; therefore, N. The fifth letter is D (Do). The crew did something about the change — they rejected the offer to return to Springfield, but it was totally wrong; therefore, N. The last letter is E (Evaluate). In this case, we also, do not have enough information to know

DECIDE MODEL CHART OF AIR ILLINOIS ACCIDENT							
Change	D	E	C	I	D	Action	E
Left generator fails after takeoff	Y	Y	N	N	N	Copilot misidentifies failed generator, disconnects good generator	Y
Copilot tells departure control of "slight" electrical problem	Y	Y	Y	Y	Y	Departure control offers return to Springfield Airport	Y
Crew gets departure control offer to return to Springfield	Y	Y	N	N	N	Captain rejects offer and continues to Carbondale	Y
Right generator doesn't take electrical load	Y	Y	Y	Y	Y	Copilot tells Captain of loss of right generator	Y
Copilot tells Captain of generator failure	Y	Y	N	N	Y	Captain requests lower altitude for VFR conditions	Y
Copilot tells Captain battery voltage dropping fast	Y	N	N	N	N	Captain tells Copilot to put load shedding switch off	Y
Copilot reminds Captain of IFR weather at Carbondale	Y	N	N	N	N	No reaction	N
Copilot turns on radar to get position	Y	Y	N	N	Y	Copilot tells Captain about dropping voltage	Y
Copilot tells Captain battery is dropping	Y	Y	Y	Y	Y	Captain turns off the radar	Y
Copilot warns Captain about low battery	Y	Y	Y	Y	Y	Captain starts descent	Y
Cockpit instruments start to fail	Y	Y	Y	N	Y	Captain asks Copilot if he's got any instruments	Y

Figure 8-1

for certain whether they did a thorough evaluation of their action. However, we give them the benefit of the doubt. Later actions perhaps would suggest differently.

Each line of the scenario can be analyzed as we have just done for the third line. It is not always possible to know exactly what the crew were thinking for each step. By attempting to determine what they may have been thinking, however, we can identify some of the thought patterns that led to the accident. After going through several exercises like this and trying to imagine what was going through the pilot's head, a process of change will begin to occur in your own thinking about how you should think when faced with similar circumstances. You will find yourself thinking about the safe outcome rather than the need to get to your destination.

A number of N's are found in the "I" column indicating there was a failure of the crew to **identify** the correct action to counter the change. However, the crucial N's occurred when the copilot reminded the captain of IFR weather at Carbondale and received no response until it was too late. The copilot appeared to have the answer to avoiding the accident but did not offer it to the captain, nor did he voice his concerns very assertively about the action the captain was taking.

The DECIDE model can be used to analyze the thought processes taking place in any accident in which sufficient communication and other behavioral data are available to answer the questions in the chart. Through such analysis the reasons behind many "pilot error" accidents might be found. From your perspective, however, the use of this technique through several scenarios, whether or not they resulted in an accident, is an effective way to learn to structure your thought processes and to make better decisions when faced with uncertain situations.

ATTITUDE

The second part of the judgment model is where the 'human element' comes into play. In this part, we see less than purely rational reasons given for the decisions that are made. Each of us develops strategies over the years to best accomplish our goals of dealing with life and the people around us.

Some of these strategies become deeply ingrained and are known as personality traits. These traits are well established by the age of six and are difficult to change thereafter. We can be quite sure that personality factors can influence a person's flying judgment. However, because of their nature, it would be futile to attempt to change them in pilot training. Furthermore, most aviation psychologists believe that personality traits are important only in the most extreme cases and can be handled through appropriate selection methods, at least for airline or military crew positions.

Attitudes, on the other hand, are strategies less deeply ingrained, which can be changed, especially under pressure from several sources at the same time. We are constantly bombarded with attempts to change our attitudes by teachers, theologians, advertising people, parents, peers, and superiors. In the cockpit, attitudes toward risk taking, as well as performance of all other aspects of flying can be modified through training.

Attitudes are also affected by nonsafety factors such as job demands, convenience, monetary gain, self-esteem, and commitment. If properly developed, this part of pilot decision-making would minimize information unrelated to the safety of the flight and direct the pilot's decision toward the use of more rational information and processes. Motivational decision-making means recognizing that hazardous attitudes are present in every human decision and that these hazardous attitudes should give way to rational thought processes.

Attitudes toward risk taking in aviation are developed within the pilot as a result of flying training and experience. Attitudes may be either positive, tending to make us take a safe or cautious approach to flying risks; or negative, causing us to take greater risks than we should.

There is no logical, rational reason why pilots should allow background forces to influence their decisions. However, these forces are a part of human nature and, therefore, cannot be discounted. They are, after all, one of the major reasons why we fly. Power, speed, adventure, and image are motivations to learn to fly, and they stay with us after we become professionals. Good judgment says we should make these forces subservient to our more rational thought processes.

Case Study

The following story from an NTSB report illustrates how far two pilots were willing to go to continue a flight because of a commitment to reach their destination, contrary to all rational safety procedures.

Two young, male, college students, both with instrument ratings and aspirations for an airline career, flew a Cessna 172 to Massachusetts for a weekend. On Sunday afternoon, they were returning to Columbus, Ohio when they made a stop for fuel in Williamsport, Pennsylvania. During their descent and approach, they ran into icing conditions but managed to land safely with about three-quarters of an inch of ice on the plane. They asked the FBO to fuel the airplane while they talked to the people in the FAA Flight Service Station (FSS) on the field about their plans to continue the next leg of their flight to Columbus.

The FSS people were astonished to hear they planned to fly back into the icing. There were numerous witnesses who talked with them and filed reports after the accident.

The Fixed-Base Operator:

"Weather was low ceilings, rain and reported icing from the surface to 6,000 feet. At approximately 1400 hours, I observed a Cessna 172 land and taxi to the fuel island at our facility. Through personal observation, ice was seen on the wings, windshield and tail surfaces.

"The pilot refueled and remained at our facility for approximately two hours. Through conversation, the pilot indicated that he and his passenger were going to proceed to Ohio State University.

"The pilot was then informed by myself and numerous other pilots in our facility that to attempt the trip with existing conditions was definitely not advisable.

"The pilot indicated that he was confident he could climb to the 6,000-foot level to get out of the ice. He then departed at approximately 1620 hours."

Pilot of a Piper Arrow:

"I had just arrived in a PARO. I was scheduled to fly a PARO to PGV but canceled the flight because of the freezing rain and icing conditions. I was explaining this cancellation to another person when one the pilots (of the C172) anxiously asked what it was like up there. I told him I was scheduled for another flight but just rescheduled it because of icing.

"He then started to knock the ice off the leading edges of the wings, struts, wheel pants, antenna and engine inlet of his airplane (172). He did not remove the streaks of ice from back of the leading edge."

State Police Helicopter Pilot:

"I left IPT at 1230 westbound, encountered ice at 500 feet AGL over the city and returned to the airport.

"I saw N1498U land with ice on leading edges and windshield. Aircraft was parked out in the freezing rain for about 1-1/2 hours. I saw the pilot beating on the leading edges of the wings, struts, and stabilator trying to remove the ice (He had moved it into an unheated hangar). I tried to remove ice from the left side stabilizer (to see how hard it was to do). I struck it with my fist and the ice did not break up. Surface was rough ice. Ice looked about 1/2 inch thick over the entire stabilizer. I saw both pilots enter N1498U, start the engine and taxi out. Ice covered stabilizer as I watched it leave the ramp."

KingAir Pilot:

"I returned from PHL in a Beech KingAir C-90 when I reached PIX and started my approach. The aircraft was covered completely with clear ice. After landing, I went to Lycoming Air's office and talked to two pilots getting ready to leave in a Cessna. I told them of the ice and told them they were 'crazy' if they were going to leave in a single-engine aircraft with no de-icing systems on the aircraft."

Radio Communications:

The following is an edited transcript of the radio communications.

2125:30   98U reported to New York Center that they were at 3,600 feet going up to six.

2128:03   98U reported that they were climbing very slowly due to ice.

2131:54   98U reported, "We're at 3,400 feet and we need to return to radar vectors (possibly to) Williamsport, unable to climb due to icing conditions."

2132:02   Center told 98U to turn left and maintain 4,000 feet direct Williamsport.

2132:09   98U reported, "Unable to maintain four thousand, the best we can do is 3,400 feet."

2138:28   98U reported its altitude was 2,500
           feet, and that was the absolute best it
           could do.

Just after 2140, 98U went off the radar screen.

The result was a crash into the trees on top of a hill near
Williamsport. Neither pilot was seriously injured. They spent
a cold night in the plane and were found by hunters the next
day.

Can you picture yourself in the position of these pilots? They
were committed to getting back to Columbus for classes the
next day. There may have been other reasons as well that we
do not know. But, whatever they were, the reasons were
sufficiently strong enough for the pilots to override the
rational decision process and to attempt a flight into conditions that were obviously extremely dangerous. Even after
expressing that they were climbing slowly at 2128:03, they
still did not turn back. Finally, at 2131:54, they gave their
first request for help — and then were still not sure they
needed to return to Williamsport, although they did follow
the ATC instructions to do so.

HAZARDOUS ATTITUDES

What would cause pilots to behave so irrationally? How can
we, as pilots, learn to prevent ourselves from making such
decisions? Five hazardous attitudes have been identified that
help to explain much of this irrational behavior. Your
awareness of these attitudes, which are found to some extent
in everyone, will help you to develop a more positive and
rational approach toward your flying decisions. These are:

Anti-Authority: This attitude is found in people who do not
                like anyone telling them what to do. They
                would say, "I don't need to follow the regu-
                lations," or "You can't tell me what to do."

Impulsivity: This attitude is found in people who fre-
                quently feel the need to do something,
                anything, immediately. They would say to
                themselves, "Do something — NOW."

Invulnerability: This attitude is found in people who believe that accidents always happen to others, not to them. They would say, "It won't happen to me."

Macho: This attitude is found in people who always try to show how good they are. They think, "I'll show you, I can do it."

Resignation: This attitude is found in people who feel they have no control over their fate. They would say, "What's the use." They are likely to go along with unreasonable requests just to be a "nice guy."

In the icing accident scenario above, these pilots had all five hazardous attitudes to some extent. They certainly had a strong sense of invulnerability. They did not think the ice would affect them as it had affected the other pilots. They also had an extreme anti-authority attitude: the regulation about not flying in known icing conditions was not for them. They also were not to be told what to do. They appeared to have a major macho attitude in that they wanted to show that they could succeed. Finally, they were impulsive in that they could not wait for the weather to improve.

In addition, these pilots had a strong case of "Get-There-Itis." This affliction has led to many aviation accidents. It occurs when a pilot has made a commitment either to himself or to others to be somewhere at a given time. In our society, because most other forms of transportation operate in all types of weather, we have a well established mental expectation to complete any and every trip we take, more or less on time. It is often difficult to avoid the same mentality in private aviation, even though we know that safe completion is highly dependent on the weather. When our ego, or self-concept, is tied up in completing our flying commitments, we have the sometimes fatal illness known to all in aviation as "Get-There-Itis."

Earlier in the chapter we told the story of the pilot who chose to drive to Victoria Falls to buy fuel, rather than fly there. That pilot was Dr. Stan Trollip, first author of this book. The following is a discussion of the thought patterns that Dr. Trollip recalls in making his decision. It illustrates how far our decision-making process can take us from the real problem at hand because of our focus on the "background problem."

"My first reaction was irritation — that this big airport could have no fuel. Fortunately I realized that being angry would not help the situation, so I started asking about other potential sources of fuel locally. I knew in my heart of hearts that I should not fly to Victoria Falls, but I felt a great deal of pressure because one of my passengers was a doctor in private practice, who had a full roster of patients the following day — several thousand dollars of income. Another passenger had meetings that he had rescheduled once already.

"I first tried to find the owners of the few other planes at the airport in the hope I could buy 30 minutes of fuel from them. Nobody knew where the owners were. I then called the few locals who owned planes to see if they had a few extra liters they would part with. No luck. I spoke to the head of the Hwange Game Park. He had fuel, but government regulations would not permit him to sell or loan any to me. My only source was Victoria Falls.

"I took out my chart and carefully remeasured the distance between Hwange and Victoria Falls and estimated that if I had an undelayed takeoff and a clear flight I would need 32 minutes of fuel. If the wind held, it may even be a little less. The fuel gauges indicated between 40 and 45 minutes of fuel. My first thought was '13 minutes leeway.' I was trying to justify a decision to go. I wanted to get my friends back on time.

"I said to myself, 'If you leave the others at the airport and go yourself, it will make the decision easier.' I measured the route again and rechecked the fuel indicators. I then realized that I was reading the needles a little high to give myself some comfort. I could see myself telling the investigating board that the gauges indicated *at least* 45 minutes, and how was I to know that they were a little inaccurate. I was already preparing my defense of a bad decision — an ominous sign! I then also realized that, in preparing my defense, I was making the assumption that I would be alive to present it (invulnerability). I started to feel less sure.

"Fortunately my friends did not put any pressure on me, because that would have been hard to resist. The turning point came when I determined what would happen if I did run out of fuel, even if I survived. I would have to spend weeks sorting out the paperwork with the Zimbabwean Government. I would be charged with flying without legal fuel reserves, and there was a good chance the plane would be confiscated, if it was not destroyed in the forced landing.

"In retrospect, it is amazing to me that I needed justifications like this to make the decision. Clearly I was still feeling completely invulnerable — confident that even if I crashed I would not be hurt. The incident was very sobering to me because I thought I was pretty well up on this judgment issue — only to find how easy it was to fall into its many traps."

Exercise 3 at the end of this chapter gives you a chance to reflect on the feelings you have had when considering a difficult decision that you have faced some time in the past. As a result of doing this type of exercise, many people have reported finding revealing thought patterns that cause them to rethink their own judgment.

A CHECK FOR HAZARDOUS ATTITUDES

At the end of this chapter, there is also a questionnaire called the "Pilot Decisional Attributes Questionnaire." This questionnaire contains 10 situations with five responses to each and is designed to show the relative amount of each of the five hazardous attitudes in your own thinking. It is **not** a test of your judgment. We all have these attitudes to some extent. This questionnaire simply shows which one may be more important to you than the others. For that reason, it may be useful to help you see which may be worth looking at when you are faced with a difficult decision in aviation.

OTHER PSYCHOLOGICAL TRAPS

There are a number of other psychological traps that pilots can fall into. You should be aware of them as well. These include, being tentative rather than decisive, "scud running," continuing VFR into IMC, losing situational awareness, casually neglecting flight planning, and being over confident. Although each of these is a behavior, all have psychological reasons for their existence. Let us consider each briefly.

THE TENTATIVE PILOT

By tentative we mean behaving as if you are uncertain what to do or unwilling or afraid to do what you know you should. It is a term from sport, especially tennis. The second author recalls when transitioning to the tailwheel airplane, he found himself in a very upsetting landing situation with a crosswind because he was being tentative in a "three-point landing." The next day his instructor demonstrated that one could force the plane onto the main wheels and hold it on the ground as one gradually lost airspeed in a "wheel landing."

SCUD RUNNING

Scud running is a term used by pilots to refer to a flight close to the ground to avoid low clouds, usually in poor visibility. Pilots get "trapped" into this activity by being too lazy to file IFR; or, if they are only a VFR pilot, they will not wait for the weather to improve.

CONTINUED VFR INTO IMC

This is the case of a VFR pilot or an IFR pilot on a VFR flight who plans the flight for VFR but, as the flight progresses, the weather deteriorates into IMC. The pilot usually thinks, "If I can just get beyond the next cloud, it will be VFR." He is thinking of the inconvenience and cost of landing and waiting out the weather. The IFR pilot may think of the inconvenience it may be to file IFR from the air or the difficulty in attracting the attention of the air traffic controller. Nevertheless, this practice has led to more fatalities than any other single factor in general aviation.

LOSS OF SITUATIONAL AWARENESS

Mental discipline is necessary while flying to keep yourself aware of all factors around you. You need to keep track of your position, which is done by using all your navigation equipment as well as by pilotage. You must always be vigilant for traffic, which requires constant visual scanning and monitoring of the radio. You need to keep abreast of the weather, which is accomplished by obtaining periodic updates throughout the flight. You must monitor your fuel consumption to ensure you can reach your destination safely. And, you need to pay attention to your passengers and crew, assuring their comfort and safety.

The moment you lose track of any one of these, you have diminished your situational awareness. This makes you more vulnerable to mistakes, particularly if a situation arises in which you feel pressure or stress.

CASUAL NEGLECT OF FLIGHT PLANNING

As pilots become more experienced in flying, there is often a tendency to gradually reduce the effort put into flight planning. Pilots who fly all-weather aircraft may even go so far as to say, "We're going anyway, why check the weather?" Proper preparation for flight takes time and is often redundant. This can tempt pilots to think, "Let's get in the air and check the weather there; if needed we can even file in the air." This added workload in the air can lead to loss of situational awareness resulting in lack of proper vigilance for other traffic.

OVER CONFIDENCE

It appears that the most vulnerable times in a pilot's career are between 100 and 200 hours of flight time. We believe the reason for this is that the pilot has become over confident. With 50 to 100 hours of flying after receiving the private license, it is easy to believe that you know all there is to know about flying. With this attitude, you relax your discipline and start glossing over details. If a problem occurs, whether due to your neglect or not, you are usually not prepared for it, increasing the likelihood of an accident.

COUNTERING HAZARDOUS ATTITUDES

We have now identified five hazardous attitudes in pilots as well as several psychological traps into which pilots can fall. Being aware of these is the first and most important factor in countering them. In addition, there are some helpful hints you can use as antidotes to these attitudes. Figure 8-2 contains a list of the five hazardous attitudes and their respective antidotes.

OTHER WAYS TO IMPROVE JUDGMENT

Most of us realize the extent to which we are prone to the five hazardous attitudes. One of the best ways to counter our susceptibility to them is to anticipate and avoid those situations that would leave us vulnerable. It is good judgment for a pilot to recognize early signs of impending trouble and to take corrective action before a critical situation can develop. The following true story illustrates:

THE FIVE ANTIDOTES TO HAZARDOUS ATTITUDES	
Hazardous Attitude	**Antidote**
Anti-Authority: The regulations are for someone else.	Follow the rules. They are usually right.
Impulsivity: I must act now, there's no time.	Not so fast. Think first.
Invulnerability: It won't happen to me.	It could happen to me.
Macho: I'll show you. I can do it.	Taking chances is foolish.
Resignation: What's the use?	I'm not helpless. I can make a difference.

Figure 8-2

A young pilot, who was recently hired by a large midwestern university, was flying in a light plane with his new chairman to attend a meeting at the Air Force Academy. Their destination, Colorado Springs, was socked in at less than 1/4 mile obscured while their alternate, Denver, was clear. The boss wanted to go to Colorado Springs to "give it a try" even though only one other plane had attempted the approach that day and was not successful. The new hire, not wanting to be placed in a situation on approach where, at minimums, the boss says, "I think I can see something — let's go lower," and having to over-rule him at that critical stage, instead over-ruled him en route and landed safely at the alternate.

In this example, the young, new hire sought to avoid a situation in which strong pressure would be brought upon him to continue an approach he regarded as hazardous. Furthermore, by not attempting the approach, he precluded the situation of having to make such a decision under the additional pressures of time. Rather, he chose to face the pressure at a time when all parties could be more rational, even though he risked being regarded as not being macho or not having the "right stuff."

FINALLY

Consider the Wright brothers' attitude toward risk as they approached powered flight. In 1901, two years before they succeeded, Wilbur Wright wrote in a letter to a friend:

> "All who are practically concerned with aerial navigation agree that the safety of the operator is more important to successful experimentation than any other point. The history of past investigation demonstrates that greater prudence is needed rather than greater skill. Only a madman would propose taking greater risks than the great constructors of earlier times."

> Quoted in Coombs, 1979

The success of the Wright brothers was due as much to their superior judgment as it was to their skill and intellect. At a time when others were being killed jumping off cliffs or out of balloons seeking adventure and the praise of the crowd, the Wrights were careful and methodical. They practiced for years while they perfected their machine, making over 900 glider flights away from the crowds so they could focus their attention on the task. They made all of their flights in the safety of the sand dunes near to the ground. They made no attempt to install the engine and fly until they were sure they had the skill they needed and a machine that was capable and reliable.

CONCLUSION

Good judgment can be very elusive. When you are sitting on the ground it is easy to make good decisions. But, when there are pressures on you to be somewhere at a certain time, it can be difficult to put all aspects of the flight into perspective.

The development of good judgment is also hampered by the fact that bad judgment is frequently reinforced because it does not result in a bad outcome. Often we make a bad decision, but nothing happens. This distorts our perception of the decision and tends to make us regard a poor decision as not being so bad. The next time we are faced with a similar situation, we are more likely to take the same risk, thinking that, because nothing bad happened before, nothing will happen this time.

It is important to remember, therefore, that the outcome does not dictate whether a decision was good or bad. A decision is good or bad irrespective of the outcome.

In the chapters to this point, we have discussed all the types of information and misinformation that pilots have to deal with. Good judgment is the ability to distinguish between these, to weigh the good with the bad, to set aside emotional pressures, and to make a rational decision that places safety as the primary consideration. Good judgment takes constant discipline, but the effort is worth it since this is the only way to enhance safety.

■ Transcript

The following is a transcript of the cockpit voice recording from Air Illinois Flight 710 that crashed on October 11, 1983 on a flight from Springfield, Illinois to Carbondale, Illinois. The recording begins at about 1 minute and 30 seconds into the flight. Prior to the start of the recording, the left generator failed, the copilot misidentified the failed generator, and disconnected the good generator. Springfield Departures Control offered the flight to return to Springfield Airport, but the captain rejected the offer and continued to Carbondale. (**CAP** - Captain, **FO** - First Officer, **DEP** - Departures Control, **KCC** - Kansas City Center, **ATT** - Flight Attendant)

TIME: 2023:54

FO: Well the left one is totally dead, the right one is putting out voltage but I can't get a load on it.

CAP: Well okay, Frank.

FO: What are we going to do?

CAP: Ah, let's concentrate on the inside here . . .

FO: I got the switch on.

CAP: What did you do, anything?

FO: Naw, reset the RCCBs (reverse current circuit breakers), I tried to select each side, isolate the side.

CAP: Yeah.

FO: Zero voltage and amps on the left side. The right one is putting out 27.5 but I can't get it to come on the line.

CAP: Okay.

DEP: Illinois 710, contact Kansas City Center 124.3

FO: *24.3 Good night.*

DEP: *Good night.*

FO: Ah, battery power is going down pretty fast . . . . *Kansas City, Illinois 710, 3,000.*

KCC: *Illinois 710, Kansas City Center, roger.*

TIME: 2026:21

FO: Ah, ya got, ah 22 volts. There's the right one.

CAP: Okay, Ah, turn the load shedding back on so they can use the reading back there, and turn off the lights, main lights.

FO: Turn off the wing lights, is that what you said? I didn't hear you.

CAP: Naw, I said I turned the load shedding switch back on so . . .

FO: Oh.

CAP: She can use the reading lights only back there.

FO: Okay. Still working with Kansas City here.

CAP: All right, I'll talk to the man here. *Kansas City, Illinois 710.*

KCC: *Illinois 710, go ahead.*

CAP: *Ah, we are kinda having a unusual request here, ah, we would like to go to 2,000 feet and if we have to go VFR that's fine, but, ah, like to, ah, like you to keep your eye on us if you can.*

KCC: *Illinois 710, I can't clear you down to 2,000, I don't even think I can keep you on radar if I had to, if you went down that far.*

CAP: *All right, fine, thank you . . .* How are our bats there?

FO: Ah, ah, 22.5.

CAP: Okay, . . . Beacons off.

FO: Okay.

CAP: Nav lights are off. Are you using these lights here?

FO: I'll get that one down.

CAP: Both generator failures, see, here . . .

FO: I'm going to try something here, I'm going to try to isolate both sides and see what happens.

(Sound of Switches)

FO: Want me to go to emergency so you can get some — get your Grimes lights?

CAP: No, I want it back the way it was. If it does reset. You see, you're shuttin' off all the electricity to the back end that way, lighting and everything.

FO: Yeah.

CAP: Allright.

FO: You want me to leave it the way it is then?

CAP: Yeah, yeah that will be good. Keep an eye on these boost pumps though.

FO: Okay

TIME: 2030:52

FO:	Are you going to try to do it separately?
CAP:	No, I — just leave it the way they are Frank. They'll be fine.
FO:	Roger that. Carbondale is 2,000 over, two, light rain and fog.
CAP:	Okay.
FO:	Winds are 150 at 10.
CAP:	Okay, got it.
FO:	Do you want me to kill any pitot heat or anything?
CAP:	I would leave the pitot heat on, it will be all right.
FO:	All right.
CAP:	Unless you see that thing really depletin,' which I don't believe it is. Is it really bad, really rapidly?
FO:	No, not too bad. Those inverts take a lot of power.
CAP:	Yeah.
FO:	All I got on here is the transponder and one nav, that's all I've got on.
CAP:	Okay, swell. DME, we don't need that.
FO:	Radar's off — only got one fan on.
CAP:	Okay.
CAP:	Are you going to be able to operate all right now on what you have back there?
ATT:	(unintelligible) . . . people want to know . . . . . .
CAP:	They want to? We have a little bit of an electrical problem here, but we're going to continue to Carbondale. We had to shut off all excess lights.
ATT:	I've only got the reading lights, the front lights by the bathroom and the baggage lights, and the entrance light.
CAP:	Okay.
ATT:	And one light by the john. What time do we get there? Is that rain?
CAP:	What time did we lift off?
FO:	There about on the hour.
ATT:	Okay.
CAP:	Do you want to use the DME?
FO:	All right.

CAP: Oh, on that checklist, other than those RCCBs then, did, ah, has been reviewed then, okay?

FO: Well, let's see here, ah RCCBs port and starboard, says trip those . . . that's about it you know, just switch both of those off and attempt to reset.

CAP: Okay.

FO: Which I've already done.

TIME: 2036:03

FO: This has just not been our day, Les.

CAP: No — that's six of one, half a dozen of another. How are we doing on them volts now?

FO: Still pretty good, 20, 21.5.

CAP: All right, ah . . .

FO: (unintelligible) . . . should last to Carbondale.

CAP: I want to use this (radar) here briefly. As a matter of fact, could you, ah tune . . .

FO: Take awhile to warm up.

CAP: Tune that in for us there, ah, there was some old guys were complaining over there around Kubik, I think we're below it altogether here. *Kansas City, Illinois 710.*

KCC: *Illinois 710, contact Kansas City Center 127.7*

CAP: *Okay.* Well.

FO: It's gonna take a few minutes to warm up, I think.

CAP: Okay.

FO: I got it.

CAP: Need a, would like him to give us a vector, I mean, if he's got us okay, we want a vector direct to the marker.

FO: Okay. *Kansas City, Illinois 710, 3,000.*

KCC: *Illinois 710, Kansas City Center roger, maintain 3,000 altimeter at Scott 29.83.*

FO: *29.83, ah if you are able, vectors direct Cabbi.*

KCC: *Illinois 710, Kansas City Center, roger, present heading looks good.*

FO: *Allright, thank you.* Well, when we lost, ah, started losing the left one I reached up and hit the right RCCB trying to isolate the right side, 'cause I assumed the problem was the right side, but they both still went off.

CAP: Well, also, too when you were doing that you see I was losing my lighting here.

FO: Yeah.

CAP: And I was losing lighting in the cabin and it was going pitch dark back there. Don't want to scare the (deleted) out of the people.

FO: Yeah, that's for sure.

CAP: Hey, it's working now, that looks like Carlyle there, either that or its a (deleted) of a shadow.

FO: Yeah, that's it — we're right on course, unbelievable.

CAP: Better stay away from them shadows, Frank . . . I suspect the circuit breaker tripped in the belly.

FO: Yeah, I was thinking the same thing, somethin' popped.

CAP: Whatever you do, don't if you would, don't say anything to dispatch . . . don't say a (deleted) thing to them.

FO: Roger that.

CAP: Not nothing.

FO: You can plan on that, that's for sure. The less you tell them about anything, the better off you are.

CAP: That's right.

TIME: 2042:15

CAP: May I have the ILS for Carbondale please?

FO: Roger that.

CAP: Still doing okay up there, Frank?

FO: Say again.

CAP: You doing all right up there?

FO: Yeah, it's at, ah, 20 volts.

CAP: Turn this thing off now.

FO: Okay.

KCC: *Illinois 710, five degrees to the right until receiving Cabbi.*

FO: *710 roger, five right to Cabbi.* Well, the boost pumps (going now) . . . The localizer should at least be doing something, getting anything at all over there?

CAP: Ah, I got the needle but, ah, little bit too far away for the flag.

FO: Yeah. Want me to tune in Cabbi real quick?

CAP: Sure.

FO: To get a bearing on it.

CAP: It's not going to use that much power.

FO: Here we go.

CAP: Is that lightning off to your right side?

FO: Say again?

CAP: Most of that lightning is off to your right side, is it not?

FO: Yeah, it's on number two.

CAP: All right.

KCC: *Air Illinois 710, contact Kansas City Center on frequency 125.3*

FO: *25.3 roger, Air Illinois 710.*

KCC: *Good night.*

TIME: 2050:37

FO: I don't know if we got enough juice to get out of this.

CAP: How come (?) Squawk your radio failure.

KCC: *Illinois 710, I've lost radar contact.*

CAP: Know your radio failure code?

KCC: *710, Kansas City.*

CAP: Frank, remember your radio failure?

FO: Yeah, I got it.

CAP: Squawk . . .

FO: Yeah.

KCC: *Illinois 710, Kansas City, do you read?*

CAP: Watch my altitude, I'm going to go down to 2,400.

FO: Okay.

CAP: You got a flashlight?

FO: Yeah.

CAP: Here we go. Do you want to shine it up here.

FO: What do you need?

CAP: Be ready for it.

FO: What do you need?

CAP: Just have it in your hand if you will.

FO: Oh, ah we're losing everything down to about 13 volts.

CAP: Okay, watch my altitude Frank.

FO: Okay, . . . 2400.

CAP: Do you have any instruments?

FO: Say again?

CAP: Do you have any instruments, do you have a horizon?

(The CVR ceases to operate here).

Exercises

Pilot Decisional Attributes Questionnaire

This questionnaire was first developed by Embry-Riddle Aeronautical University under contract to the Federal Aviation Administration (Berlin, J.I., et al., 1982.) What follows is a modified version of the original one.

Instructions: The following is a questionnaire designed to assist you in determining your decisional attributes as a pilot. Please answer all of the questions as honestly as you can. There are no right or wrong answers.

1.   **Read each of the situations and the five alternatives. As you read, place yourself in a familiar aircraft under the given circumstances.**

2.   **Choose your MOST LIKELY thought pattern in response to the situation and place a rank of "1" beside it. Then rank each of the other responses in descending order of preference from 2 through 5.**

Remember, this questionnaire has no correct answers. Some (or all) of the alternatives given may not represent the way you would think at all. However, for the purposes of this questionnaire, you should assign a rank to all of them.

Situation 1

You are on a flight to an unfamiliar, rural airport. Flight service states that VFR flight is not recommended since heavy coastal fog is forecast to move into the destination airport area about the time you expect to land. You first consider returning to your home base where visibility is still good but decide instead to continue as planned, and land safely after some problems. Why did you reach this decision?

___a.   You hate to admit that you cannot complete your original flight plan.

___b.   You resent the suggestion by flight service that you should change your mind.

___c.   You feel sure that things will turn out safely, and that there is no danger.

___d.   You reason that since your actions would make no real difference you might as well continue.

___e.   You feel the need to decide quickly so you take the simplest alternative.

Situation 2

While taxiing for takeoff, you notice that your right brake pedal is softer than the left. Once airborne, you are sufficiently concerned about the problem to radio for information. Since strong winds are reported at your destination, an experienced pilot who is a passenger recommends that you abandon the flight and return to your departure airport. You choose to continue the flight and experience no further difficulties. Why did you continue?

___a.   You feel that suggestions made in this type of situation are usually overly cautious.

___b.   Your brakes have never failed before, so you doubt that they will this time.

___c.   You feel that you can leave the decision to the tower at your destination.

___d.   You immediately decide that you want to continue.

___e.   You are sure that if anyone could handle the landing, you can.

Situation 3

Your regular airplane has been grounded because of an airframe problem. You are scheduled in another airplane and discover it is a model with which you are not familiar. After your preflight, you decide to take off on your business trip as planned. What is your reasoning?

___a.   You feel that a difficult situation will not arise so there is no reason not to go.

___b.   You tell yourself that if there were any danger you would not have been offered the plane.

___c.   You are in a hurry and do not want to take the time to think of alternate choices.

___d.   You do not want to admit that you may have trouble flying an unfamiliar airplane.

___e.   You are convinced that your flight instructor was much too conservative and pessimistic when he cautioned you to be thoroughly checked out in an unfamiliar aircraft.

Situation 4

You were briefed about possible icing conditions but did not think there would be any problem since your departure airport temperature was 60°F. As you near your destination, you encounter freezing precipitation, which clings to your aircraft. Your passenger, who is a more experienced pilot, begins to panic. You consider turning back to the departure airport but continue instead. Why did you not return?

___a.   You feel that, having come this far, things are out of your hands.

___b.   The panic of the passenger makes you "commit yourself" without thinking over the situation.

___c.   You do not want the passenger to think you are afraid.

___d.   You are determined not to let the passenger think he can influence what you do.

___e.   You do not believe that the icing could cause your plane to crash in these circumstances.

Situation 5

You do not bother to check weather conditions at your destination. En route, you encounter headwinds. Your fuel supply is adequate to reach your destination, but there is almost no reserve for emergencies. You continue the flight and land with a nearly dry tank. What most influenced you to do this?

___a.   Being unhappy with the pressure of having to choose what to do, you make a snap decision.

___b.   You do not want your friends to hear that you had to turn back.

___c.   You feel that flight manuals always understate the safety margin on fuel tank capacity.

___d.   You believe that all things usually turn out well, and this will be no exception.

___e.   You reason that the situation has already been determined because the destination is closer than any other airport.

Situation 6

You are forty minutes late for a trip in a small airplane and since the aircraft handled well on the previous day's flight you decide to skip most of the preflight check. What leads you to this decision?

___a.   You simply take the first approach to making up time that comes to mind.

___b.   You feel that your reputation for being on time demands that you cut corners when necessary.

___c.   You believe that some of the preflight inspection is just a waste of time.

___d.   You see no reason to think that something unfortunate will happen during this flight.

___e.   If any problems develop, the responsibility would not be yours. It is the maintenance of the airplane that really makes the difference.

Situation 7

You are to fly an aircraft which you know is old and has been poorly maintained. A higher-than-normal r.p.m. drop on the magneto check is indicated, and you suspect the spark plugs. Your friends, who are traveling as passengers, do not want to be delayed. After five minutes of debate, you agree to make the trip. Why did you permit yourself to be persuaded?

___a.   You feel you must always prove your ability as a pilot, even under less-than-ideal circumstances.

___b.   You believe that regulations overstress safety in this kind of situation.

___c.   You think the spark plugs will certainly last for just one more flight.

___d.   You feel your opinion may be wrong since all the passengers are willing to take the risk.

___e.   The thought of changing arrangements is too annoying, so you jump at the suggestion of the passengers.

Situation 8

You are on final approach when you notice a large unidentified object on the far end of the runway. You consider going around, but your friend suggests landing anyway since the runway is "plenty long enough." You land, stopping 200 feet short of the obstacle. Why did you agree to land?

___a.   You have never had an accident, so you feel that nothing will happen this time.

___b.   You are pleased to have someone else help with the decision and decide your friend is right.

___c.   You do not have much time, so you just go ahead and act on your friend's suggestion.

___d.   You want to show your friend that you can stop the plane as quickly as needed.

___e.   You feel the regulations making the pilot responsible for the safe operation of the aircraft do not apply here since it is the airport's responsibility to maintain the runway.

Situation 9

You have just completed your base leg for a landing on runway 14 at an uncontrolled airport. As you turn to final, you see that the wind has changed, blowing from about 90°. You make two sharp turns and land on runway 11. What was your reasoning?

___a.   You believe you are a really good pilot who can safely make sudden maneuvers.

___b.   You believe your flight instructor was overly cautious when insisting that a pilot must go around rather than make sudden course changes while on final approach.

___c.   You know there would be no danger in making the sudden turns because you do things like this all the time.

___d.   You know landing into the wind is best, so you act as soon as you can to avoid a crosswind landing.

___e.   The unexpected wind change is a bad break, but you figure, if the wind can change, so can you.

Situation 10

You have flown to your destination airfield only in daylight and believe you know it well. You learn that your airplane needs a minor repair which will delay your arrival until well after dark. Although a good portion of the flight is after dark, you feel you should be able to recognize some of the lighted landmarks. Why did you decide to make the flight?

___a.   You believe that when your time comes you cannot escape, and until that time there is no need to worry.

___b.   You do not want to wait to study other options, so you carry out your first plan.

___c.   You feel that if anyone can handle this problem, you can do it.

___d.   You believe the repair is not necessary. You decide you will not let recommended but minor maintenance stop you from getting to your destination.

___e.   You simply do not believe that you could get off course despite your unfamiliarity with ground references at night.

PILOT DECISIONAL ATTRIBUTES
QUESTIONNAIRE

Scoring Key

Scoring Instructions: From your answers to each situation, write your rank for each alternative in the table below. Sum the ranking scores for each scale and enter at the bottom. These totals should then be marked on the Attitude Inventory Profile Graph.

Situation	Scale I	Scale II	Scale III	Scale IV	Scale V	Total
1.	b____	e____	c____	a____	d____	15
2.	a____	d____	b____	e____	c____	15
3.	e____	c____	a____	d____	b____	15
4.	d____	b____	e____	c____	a____	15
5.	c____	a____	d____	b____	e____	15
6.	c____	a____	d____	b____	e____	15
7.	b____	e____	c____	a____	d____	15
8.	e____	c____	a____	d____	b____	15
9.	b____	d____	c____	a____	e____	15
10.	d____	b____	e____	c____	a____	15
TOTAL	____	____	____	____	____	150

The sum of your scores across must be 15 for each situation. If it is not, go back and make sure you transferred the scores correctly and check your addition. The grand total should be 150.

ATTITUDE INVENTORY
PROFILE GRAPH

1.   Enter the raw score obtained from each scoring key in the correct blank space below. The sum of the five scores should equal 150. If it does not, go back and recheck your work.

        Scale I (Anti-authority) _____

        Scale II (Impulsivity) _____

        Scale III (Invulnerability) _____

        Scale IV (Macho) _____

        Scale V (Resignation) _____

2.   Now look at the profile graph shown on the following page. Notice that there are five vertical scales, one for each of the raw scores shown above. Place a mark on each scale corresponding to your score. Draw the lines that connect the five marks.

3.   **Profile Interpretation**. Now that you have completed your judgment profile, you may wonder what it means. First, you should know that there is no ideal profile for pilots. The profile simply shows the relative strength of one hazardous attitude over another as you face stressful decisions. Second, because you ranked the five items from 1 to 5 (most likely to least likely), the scale with the lowest score represents the strongest attitude.

You should examine your profile to see whether any of the scales are lower than the others. If so, you should examine yourself as you are faced with stressful decisions to see whether that attitude is influencing your thinking. If you think it is, try to correct it by moving your thoughts toward the more rational ideas that lead to a **safe** conclusion. Remember, everyone has all of these attitudes to some extent. The pilot with good judgment has learned to make these attitudes subservient to more rational thought processes.

ATTITUDE INVENTORY PROFILE GRAPH				
Scale I	Scale II	Scale III	Scale IV	Scale V
50	50	50	50	50
40	40	40	40	40
30	30	30	30	30
20	20	20	20	20
10	10	10	10	10
Anti-Authority	Impulsivity	Invulnerability	Macho	Resignation

The Decide Model Chart

Instructions: Construct a DECIDE model chart for the following accident scenario described in *Aviation Consumer*, May 15, 1980. You may copy the chart at the end of the story. When you have completed the chart, see how many errors in judgment you can find.

This is a story of a new private pilot whose basic flying skills never let him down; but, in our opinion, he died because of a profound lack of judgment. He was a 31-year-old western Texas oil field worker who obtained his private license October 30, 1978. He had a total of 107 hours — 58 dual and 49 solo; he had 3.9 hours on simulated instruments, 2.9 hours night dual, and 4.9 hours night solo.

On December 15, 1978, a Friday, the pilot worked a full shift in the oil fields and then had a friend fly him to Midland,

Texas, where he picked up his rental airplane. His plan was to fly to his home field in Colorado City, Texas, pick up his wife and children, and fly on to Memphis, Tennessee. The FBO at Midland told investigators he rented the pilot a Piper Archer II, but gently suggested that he might want to stay overnight at home and continue to Memphis in the morning. But the pilot said he wanted to beat a cold front to Memphis and added that, anyway, he "liked to fly at night." He departed a little after 5 p.m. with topped-off tanks.

His friend was waiting at Colorado City, and later told investigators it took the pilot three tries to land since he did so downwind. Even so, he barely got the airplane stopped before running off the end of the 3,000-foot runway.

The pilot loaded up his passengers: wife, five-year-old son, and three-year-old daughter. His friend suggested waiting until the next morning; this was rejected. The friend asked if he could top off the tanks, but the pilot said he was in a hurry and, besides, had plenty of fuel to reach Texarkana, Arkansas, where he planned to refuel. The pilot left without filing a flight plan or getting an FSS weather briefing.

The next known contact with Archer 6902J was around 8:35 p.m., when the plane is assumed to have been approaching Texarkana. The pilot called Shreveport (LA) FSS and inquired about weather at various fields in the area. Some had gone down to marginal VFR, others were IFR, and it was clear that a ceiling was developing from about 3,000 to 6,000 feet with fog rolling in underneath. The Archer was apparently on top. Part of the conversation went:

> FSS: "You are IFR, is that correct?"
> 02J: "It is a VOR, ah VFR."
> FSS: "A VFR flight plan, sir?"
> 02J: "Hey, 10-4 on a VFR."

The pilot inquired about Memphis weather and got a brighter picture there — 3,000-foot ceilings — expected to stay that way — and clear of fog. (However, our calculations indicate that he could hardly make it to Memphis with the fuel remaining.) The pilot thanked the briefer and left the frequency at about 8:45 p.m.

He was not heard from again until about 10:15 p.m., when Shreveport heard weak transmissions from an airplane "in trouble." It took some minutes while controllers and FSS staffers coordinated, trying all channels and asking for relays from airliners, before Fort Worth Center established contact with 6902J.

When identified on radar at about 10:29, the plane was still about 25 miles west of Texarkana Municipal Airport; the pilot had apparently flown in circles for well over an hour since talking to the FSS. The pilot told the controller, "I'm in trouble, I'm in the clouds, I can't see and I'm lost." The controller issued vectors for Texarkana. The pilot mentioned having about 15 minutes of fuel remaining; the controller reminded him to use his most economical fuel setting. The controller queried, "You are IFR qualified?" The pilot replied, "Negatory, Negative."

There was a handoff and at 10:34 the new controller asked "How much time you got left?" The pilot responded, "Uh, I'm . . . I'm . . . It's hard to say. It's showing empty now. I'm saying about 10 minutes of fuel left." (Archer capacity: 50 gal. total, 48 usable.)

Weather at the airport was: indefinite ceiling 200 feet, sky obscured, visibility one-half mile in fog. The lowest IFR minima at the field were 200 feet and a half-mile. Runway 22 had an ILS, but to use it the flight would have to overfly the airport and past the localizer (back course) approach to runway 4 and a VOR approach to runway 13. The controllers chose to issue vectors for the straightest possible course to runway 4.

For a time, 6902J was above a cloud deck at 5,300 feet. This conversation ensued:

Center:	"How much flying experience do you have?"
02J:	"Very little. Uh, I just got my private license and I'm still a student."
Center:	"Okay, 02J, have you had any instrument at all?"
02J:	"Uh, just on private training."

Later, the controller gave a little pep talk: "Okay, 02J, we're gonna have to go through these clouds here pretty soon now, so get used to the . . . get your head inside the cockpit if possible and get used to the artificial horizon again. It's just like you did in the training."

And so began the approach — the controller giving vectors and keeping up spirits, the young pilot complying precisely. In this manner, the plane was guided right down the runway, but descending as low as 650 feet MSL (250 AGL), the pilot could not see the runway lights. The controller was starting instructions for a missed approach, when the pilot reported, "My wife said she seen the lights." The pilot attempted to turn around and come back from north of the field. Responding to a heading instruction, he said, "Roger, let me just, up just a little bit. I'm at 400 feet and I just seen a house."

The controller (amazingly) kept his cool: "All right, 02J, at that altitude you should be below . . . there's nothing, uh, no restrictions on your approach from the north . . . uh, remember the field elevation is 399 feet."

The last conversation:

> 02J: "Roger, now give me the heading."
>
> Center: "Okay, heading 160, 160, 02J."
>
> 02J: "Roger, just a second . . . Uh, give me a heading now. I'm hurting. I'm out of gas."
>
> Center: "Okay, heading 130, heading 130. You are one mile from the airport, heading 130."

The flight did not answer. It had crashed, nose steeply down, wing low, into heavy trees. Only the girl survived, with serious injuries.

Complete the DECIDE chart on the following page and count the number of errors of judgment you can find. In this accident, we count at least 15 separate and pronounced errors in judgment.

DECIDE MODEL CHART OF TEXAS OIL FIELD WORKER ACCIDENT							
Change	D	E	C	I	D	Action	E
1.							
2.							
3.							
4.							
5.							
6.							
7.							
8.							
9.							
10.							
11.							
12.							
13.							
14.							
15.							
16.							
17.							
18.							
19.							
20.							

Pilot Judgment Questionnaire

Instructions: Complete the following Judgment Question-naire. There are no correct or incorrect answers. Its purpose is to make you more aware of your own strengths and weaknesses in making judgments about flying. Use the questionnaire as a vehicle for discussion with your pilot friends or instructor.

1.   Have you ever made a flying decision which, upon reflection, you would have considered crazy or bad judgment and you survived more because of luck than good sense?

     Yes _____

     No _____

2.   What types of risky (less than 100% certainty of success) choices have you faced in your occupational flying career?

Safety Factor	Number of Times Faced per Year
Weather	_____
Mechanical	_____
Electrical	_____
Aircraft Loading	_____
Aircraft Separation (ATC)	_____
Runway Length (Conditions)	_____

3. Which of the following pressures have you faced in choosing to make a flight in marginally safe conditions?

Pressure Source	**Number of Times Faced per Year**
Duty	_____
Economic	_____
Peers	_____
Adventure (Challenge)	_____
Superiors (On Job)	_____
Social (Friends)	_____

4. Briefly describe a difficult decision situation from your flying experience:

_____

_____

_____

_____

5. In the situation you described above, what were your decision alternatives?

1. _____

2. _____

3. _____

6. What were the safety factors making this a decision worth thinking about? Weather? Aircraft? People? Other? Briefly describe.

    _____

    _____

    _____

    _____

7. What pressures did you feel to make or continue the flight in spite of the safety factors? Personal commitment? Monetary? Passenger requirement? Other? Briefly describe.

    _____

    _____

    _____

8. How much time did you have to make your decision?

    _____

9. Did you have all of the information needed to make the decision?

    Yes _____

    No _____

10. What additional information could you have used?

    1. _____

    2. _____

    3. _____

11.   Indicate the level of uncertainty that you felt in making your decision. Draw an "X" at level of uncertainty.

100%	50%	0%
Certain	Certain	Pure Guess

12.   Did you get assistance from anyone in the cockpit, cabin or ATC in either gathering information or making your choice?

	Yes	No
Gathering Information	____	____
Making Choice	____	____

13.   Optional: What was your decision? Briefly describe:

_____

_____

_____

REFERENCES AND RECOMMENDED READINGS

AOPA Pilot magazine column: "Never again." Frederick, MD: Aircraft Owners and Pilots Association (AOPA).

Benner, L. 1975. D.E.C.I.D.E. in hazardous materials emergencies. *Fire Journal*, 69(4), pp. 13-18.

Berlin, J.I., Gruber, E.V., Jensen, P.K., Holmes, C.W., Lau, J.R., Mills, J.W. & O'Kane, J.M. 1982. *Pilot judgment training and evaluation: Volume 1*. Atlantic City, NJ: FAA Technical Center, Report No. DOT/FAA/CT-82/56.

Coombs, H. 1979. *Kill Devil Hill*. Boston: Houghton Mifflin Company.

Flight Instructors' Safety Report. Frederick, MD: AOPA Air Safety Foundation.

Flying magazine column: "I learned about flying from that." New York: Diamandis Communications Inc.

Hawkins, F.H. 1987. *Human Factors in Flight*. Aldershot, England: Gower Technical Press.

Jensen, R. S., Adrion, J., & Maresh, J. 1986. A preliminary investigation of the application of the DECIDE model to aeronautical decision making. Columbus, OH: The OSU Advanced Flight Simulation Laboratory, Final Report No. AOPA 86-3.

NAFI Foundation Newsletter. Dublin, OH: National Association of Flight Instructors.

NTSB Reporter. Yonkers, NY: Peter Katz Productions, Inc.

O'Hare, D. & S.N. Roscoe. 1990. *Flight Deck Performance*. Ames, IA: Iowa State University Press.

Wickens, C.D. & J.M. Flach. 1988. Information Processing. In E.L. Wiener & D.C. Nagel (Eds.) *Human Factors in Aviation*. San Diego, CA: Academic Press.

Cockpit Resource Management **9**

INTRODUCTION

"In the past year and a half, an alarming number of safety investigation boards have concluded crew error was causal. Equally alarming is the realization that often the information missing from the critical, often final, decision was available to the crew. In some cases, at least one crew member had the answer. While mishaps are always tragic, those in which the resources to prevent catastrophe were available and either unrecognized, unused, or simply not offered represent an especially intolerable category."

MAC Flyer, 1985

Over the last few years, there has been an alarming number of airline accidents in which faulty Cockpit Resource Management (CRM) has been cited as a factor. Statistics show, in fact, that over half of the accidents in the airline industry are the result of a faulty application of CRM. Four accidents, in particular, have been prime motivators for the development of training programs to teach or improve CRM skills. These are:

*   Eastern, L-1011, Miami, 1972 — the aircraft descended into the Everglades while the crew was distracted trying to resolve a landing gear problem that turned out to be a burned-out bulb.

*   Pan American and KLM, both B-747s, at Tenerife, Canary Islands, 1977 — during the takeoff roll, KLM collided with Pan American due to a mix-up in communications.

*   United, DC-8, Portland, 1978 — the aircraft ran out of fuel and crashed short of the runway due to distractions and a breakdown in cockpit communications about the fuel state.

*   Air Florida, B-737, Washington National, 1982 — the aircraft crashed shortly after takeoff due to airframe and engine icing. The co-pilot expressed concern prior to and during the takeoff that all was not right.

In each of these accidents, the captain failed to make effective decisions because he, or his crew, did not use proper CRM practices.

BASIC CRM CONCEPT

Cockpit Resource Management refers to the effective use of all resources to achieve safe and efficient flight operations and is an extension of pilot judgment to the multi-person flight crew. The primary focus of CRM is on communication, both among those who are inside the cockpit and those who are outside, such as ATC. Communication is the critical factor in cockpit management because it is the means by which management is carried out.

CRM is generally regarded as a topic of discussion for multi-person crews, and many of the issues we discuss certainly relate better to them than to a single-pilot situation. However, we have included this CRM chapter, because it has a direct application to general aviation as well. You may wonder how this is so, since relatively few general aviation flights require more than a single crew member. In reality, a large number of general aviation pilots do not make their flights solo, but have a friend, instructor, or another pilot in the right seat. In such cases, CRM training can be helpful to improve effective cockpit management and safety. Many general aviation pilots also anticipate moving on to larger airplanes that require multi-person crews; therefore, CRM training early in their flight education helps to establish teamwork and communication skills at a point in training when they can be more effectively learned.

ASSUMPTIONS

Before considering CRM training, two assumptions are made about you. The first is that you have technical flying competence (stick and rudder skills), the cornerstone of effective pilot performance. Thus, there is no need to discuss the skills of flying the airplane at all. Second, although personality does affect crew management performance, there is no attempt to change your personality in CRM training.

CRM training is designed to address behavior as a product of all the things that make you who you are. Your knowledge, way of thinking, personality, attitude, and experience all influence your behavior. Although CRM training does not focus on personality change, it can teach you ways to think clearly in group decision making and can have an impact on your attitude, interpersonal communication, leadership, and reaction to stress. It can help you be more flexible in how you handle difficult situations, as well as improve your ability to maximize resources in critical situations when effectiveness is a life-or-death issue.

GOALS OF CRM TRAINING

The primary goal of CRM training is accurate, effective aeronautical decision making. The key to good cockpit management is communication among crew members.

Information must be requested, offered, and/or given freely in a timely way to permit accurate, sound decisions. To accomplish this, you have to develop an effective interpersonal communication style, leadership skills, and decision making skills. You also have to develop a "Team Concept" and learn to deal with stress.

Dealing with stress is an important component of CRM because, as we discussed in the chapter on Emotional Stress, one of its most significant and universal effects is a reduction in verbal communication.

PERSONALITY VERSUS ATTITUDE

Personality and attitude are among the factors known to affect CRM performance. The term **personality** refers to the relatively enduring characteristics you acquired during your developmental years. Over the years of development, each of us develops strategies to best accomplish our goals of dealing with life and the people around us. Some of these strategies become deeply ingrained and are known as personality traits, which are well established by the age of six. These traits are quite difficult to change and are modifiable only through the considerable efforts of psychotherapy. Because of their nature, it would be futile to attempt to change personality traits in pilot training, and no such suggestion is made in CRM training. Furthermore, most aviation psychologists believe that only extremes of personality are likely to affect safety, and that people with such personalities are not likely to be in the cockpit in the first place.

On the other hand, **attitudes** are less deeply internalized components of your personality and are subject to change fairly easily, especially under pressure from several sources at the same time. Attitudes are constantly bombarded by forces in our society such as advertisers, salesmen, teachers, politicians, and preachers. In the cockpit, management attitudes can be modified through training, as can attitudes toward risk taking and other aspects of the flying task. Because many of these attempts to change attitudes have been successful in other fields, we know that they can be used in cockpit training as well. Therefore, a major thrust of CRM training is to improve the attitudes of the flight crew so as to bring about better team decision making.

RELATIONSHIPS VERSUS TASKS

Two aspects of attitude that are very important to CRM are your concern for relationships with others and your concern for completing the task at hand. The relationship versus task model is usually presented as a matrix with one side representing the **Relationship** orientation and the other side representing the **Task** orientation. As shown in figure 9-1, there are four quadrants in this model representing the four behavioral styles: nurturing, autonomous, assertive, and aggressive. To find out where you fall, complete the Personal Characteristics Inventory at the end of the chapter.

According to this model, your behavioral orientation, including communication style, can be described by where you fall on the relationship and task scales. People who have a high relationship orientation tend to consider the feelings of others first, believing that the task can be best accomplished when everyone gets along well with each other. On the other hand, those who have a high task orientation tend to consider the accomplishment of the task more important than relationships and would act to get the job done whatever the cost to relationships. The ideal orientation for flight crews in most flight situations is a strong combination of both.

In the application of this model to your cockpit, the assertive style of behavior is advocated because it produces best

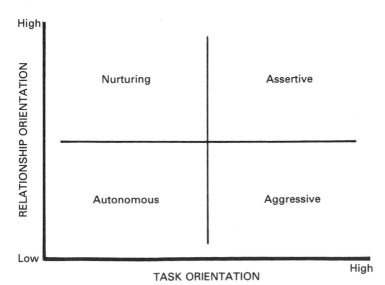

Figure 9-1

overall performance in decision making. Although you may come to the cockpit with certain tendencies to be either nurturing, autonomous, and/or aggressive, you should adapt these to meet the demands of each situation, regardless of your role in the cockpit. Extremes of each type are to be avoided because they discourage effective communication.

Different situations may require variations on a basically assertive style. For example, an emergency may be handled most effectively with a combination aggressive/assertive style, because there is little time to consider relationships or gather more information. Other situations may demand a nurturing/assertive style, such as helping a crew member through a mistake without destroying his or her ego. Finally, you must be sure that the other people you are working with know how you are behaving at all times, and it is equally important that you know how the other(s) are acting as well.

As you can see from the expanded version of the relationship versus task model shown in figure 9-2, we suggest you develop flexibility to move toward one or another of the

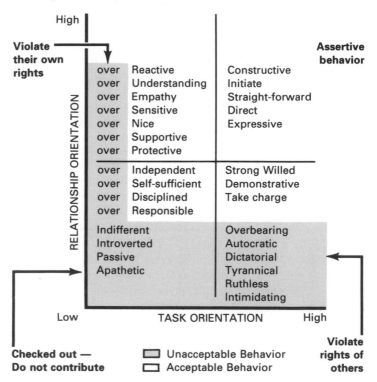

Figure 9-2

three alternate quadrants depending on the situation. Overall, you should attempt to maintain a basic assertive style. You can also see that extreme behaviors in each of the non-assertive quadrants (shaded area) can lead to failures in communication and relationships. Those who are excessive in nurturing tend to violate their own rights. Those who are too aggressive violate the rights of others. Those who place too much emphasis on their own autonomy may fail to contribute anything useful.

COMMUNICATION

The most important aspect of CRM training is interpersonal communication. It is through communication that management is conducted, and it is the responsibility of all crew members to communicate effectively. Good communication means more than speaking clearly with proper phraseology. It also means ensuring that the other person understands what you are saying, as well as you understand what the other person is saying. It is helpful to recognize your own communication style and those of your colleagues, and it is important to keep improving your own communication.

There are five aspects of communication on which to focus. Improving your proficiency in inquiry, advocacy, listening, conflict resolution, and critique can help you become a good resource manager and decision maker. As you can see, none of these deal with the details of proper grammar or clear meaning. All deal with the transfer of information important to understanding.

Because of the importance of each of these aspects of communication, we discuss each in some detail with cockpit examples. Later, we will present case studies to further your understanding of these ideas. It is important that you not only understand the concepts presented here, but also that you adapt them to your own communication practices.

INQUIRY

Inquiry or information-seeking is the first aspect of communication covered, because it represents the beginning point of making effective decisions. Good decisions are based on good information. In the cockpit, inquiry consists of gathering information both from your visual scan and from questioning other crew members or outside sources, such as air traffic control. It also means asking for clarification when the information is not clearly understood. It also means checking your instruments, radios, charts, and visually scanning for aircraft outside of the cockpit.

In the cockpit, people with fragile egos are often reluctant to ask for clarification, because they think it may reflect badly on their intellect or hearing. The situation can be made worse when other equally insecure peers ridicule others for not understanding what was said. This feeling of an insecure ego must be overcome if complete understanding is to be achieved and safe decisions are to be made consistently.

A good example of the problems that can occur during the inquiry process is the Air Florida accident. The co-pilot, who was flying, asked the captain, "Slushy runway, do you want me to (do) anything special for this or just go for it?" This was a clear inquiry about ideas for the takeoff, as well as an unstated expression of concern for the takeoff. The captain's response was to ridicule the question: "Unless you got anything special you'd like to do." A better response would have been to ask the co-pilot what he was concerned about. Following the captain's response, the co-pilot then repeated his concern three more times in different, less direct, ways without a response from the captain.

So, the first step towards good communication is not to be afraid to ask questions.

ADVOCACY

Advocacy refers to the need to state what you know or believe in a forthright manner. It means not only stating your position, but maintaining your position until completely convinced by the facts, not the authority of the other person, that it is wrong. There are many examples of airline accidents in which other crew members may have had the correct answer, as indicated by the questioning of the actions of the captain, but did not advocate their position strongly enough or capitulated to the authority of the captain far too soon. Two examples of this are the United accident in Portland in which the flight engineer expressed concern over the fuel situation, and the Air Florida accident in Washington in which the co-pilot indicated concern for the difference in the engine temperatures. In both cases, the captain's decision would have benefitted if the subordinate crew member had been a more tenacious advocate of his position.

The second step, then, towards improving communication is to make your thoughts and feelings known.

LISTENING

Active listening is an important part of successful CRM. One of the greatest reasons for cockpit communication failures is the fact that no one was listening. Listening requires more than passive attention. It requires you to open up to the other person, actively inquire through questions and other forms of feedback, and respond appropriately by agreeing, acknowledging, or disagreeing. Listening is not passive — it is a part of communication for which everyone is responsible. No one must behave like a sponge, absorbing without giving.

You are likely not a good listener if you have the following traits:

Pre-plan: You are so intent on what you want to say that you do not listen to what others are saying.

Debate: No matter what is said, you want to take the other side, "I'm going to play the devil's advocate."

Detour: This is similar to pre-plan, but you wait for a key word to take the discussion to another area of interest to you.

Tune out: Whatever the other person says, it is not important enough for your attention.

On the other hand, you are a good listener if you have the following traits:

Ask questions: You seek clarification or elaboration.

Paraphrase: You restate the speaker's position.

Make eye contact: You make eye contact to put the other person at ease.

Use positive body language: Your body language invites further communication.

Good listening results in better communication, safety, and efficiency. It also promotes relationships, improves decision making, and creates harmony. To summarize active listening, the following points are offered:

1. It is a basic human need to be heard and understood —
    active listening serves that need.
2. Good listening is a skill that must be learned.
3. In an emergency, good listening is a critical skill.
4. In normal situations, good listening enhances commu-
    nication, eliminates barriers, and lays the groundwork
    for good communication during emergencies.

The third step towards good communication, then, is to
become a good listener.

CONFLICT RESOLUTION

If each person in the cockpit advocates his or her respective
position properly, conflict is inevitable. Therefore, an
effective process is needed to resolve those conflicts. Conflicts are not necessarily bad as long as they arise over issues
within the cockpit. They can become destructive when issues
from outside the cockpit are brought into the argument, such
as taking sides on management policies, personality factors,
personal weaknesses, social status, and so on. It can also be
destructive when the argument is over who is right rather
than what is right. Such arguments can have a serious effect
on the quality of the decisions made, because they focus your
attention on irrelevant issues.

On the other hand, conflict can be very constructive if it is
handled properly. When you are pilot-in-command, using the
following practices can help resolve conflicts with positive
results:

1. Encourage others to express their opinions.
2. When disagreement arises, keep the discussion on the
    issues needing resolution within the cockpit.
3. Bring out all issues of disagreement.
4. Acknowledge and express all feelings that are deep
    enough to cloud your thinking. Your colleagues should
    know why you feel so strongly about an issue.

Properly handled, conflict resolution is fundamental to good
problem-solving. It leads to deeper thinking, creative new
ideas, mutual respect, and higher self-esteem, all of which

strengthen team effectiveness. For these reasons, conflict should not be avoided when differences of opinion arise. Rather, it should be recognized as an opportunity to seek better solutions that may not have been thought of previously.

The fourth step towards good communication is to improve your ability to resolve or prevent conflicts in the cockpit.

CRITIQUE

Even more difficult than conflict resolution is the ability to provide an effective critique of fellow crew members. No one is perfect and mistakes are likely to occur on every flight. A good critique can help the person who made the mistake improve his or her future performance. Your instruments provide feedback about your flying skills, such as tracking. However, if you want to improve other cockpit skills, such as problem-solving, monitoring traffic, communicating, and so on, you need feedback in the form of a critique from your colleagues in the cockpit.

The problem is who is responsible for giving the critique. Because of his or her position, the captain or pilot-in-command has first responsibility for providing feedback. However, this person also makes mistakes that need to be pointed out. This is a much more difficult situation, because social norms do not encourage you to question those in authority over you. In a multi-person crew, you may feel concerned that the captain will censure you or, worse still, place roadblocks in your career path. Nevertheless, captains also need feedback, and often lament the fact that they have never received any feedback concerning their performance except on proficiency checks.

The question is how to provide a critique? First, all members of the crew should know to expect a critique. CRM training is useful in bringing about this awareness. Second, to reduce pain and embarrassment, it is good practice to ask for feedback, rather than waiting for it to be given. This is especially true for the captain. Third, the critique should consist of on-going, frank discussions among the crew members. This should start at the beginning of the flight planning, continue through the flight, and end with a debriefing at the conclusion of the flight. If properly done, it can become a way of life in the cockpit to resolve conflict and misunderstandings before they arise. A good critique can also prevent issues and important feelings from being covered up.

The attitude and reaction of the person receiving the critique may be just as important as the initiative of the person giving it. Your attitude should encourage the other to give feedback. On the other hand, you should realize that feedback from any one person cannot be fully trusted, because it comes from one perspective. Everyone who sees you perform, sees it in relation to their own experience and background, which influences the feedback they give you. Therefore, it is necessary to get feedback from others to get an accurate picture of your performance.

So, the final step toward good communication is developing the ability to give and receive constructive criticism.

LEADERSHIP

At the heart of cockpit resource management is effective leadership. Each member of the crew must recognize that he or she has a leadership responsibility that is important to effective decision making. No matter which position you occupy in the cockpit, you must learn to become a leader in that position.

The following are concepts generally acknowledged to relate to leadership. You will notice that many of these traits are the same as those given above for effective CRM performance.

Competence: A cockpit leader is first and foremost a competent pilot. Piloting skills must be exemplary and should inspire the confidence of other crew members. Co-pilots and flight engineers must have a mastery of their job skills, which complements the captain. If you are a single-pilot, the more competent you are, the more others will respect your position and opinion. Others can sense competence, and the competent pilot will frequently be given preferential treatment by ATC or line personnel.

Communication skills: Leaders tend to look the part and have a wide-ranging vocabulary; they inspire individual and group confidence.

Listening skills: Leaders listen. They interpret and evaluate what they hear, and do not permit personal ideas, emotions, or prejudices to distort what a person says. Disciplined listening prevents them from tuning out subjects they consider too complex or uninteresting.

Decision making: An effective leader is skillful at problem analysis and decision making and seeks out all pertinent information in arriving at a decision. It is easy to make decisions based upon narrow information, but the results are generally less than optimum. Limited information, which is readily available, sometimes presents an incomplete or misleading picture of the situation. Making an extra effort to seek out additional information may place a new perspective upon the situation.

Decisiveness: Leaders view decisiveness as an essential binding influence for unity of action. Followers will usually excuse almost any stupidity, indiscretion, or ill-conceived action, but they will not accept excessive timidity.

Perseverance: People who aspire to or have achieved leadership persevere in their work. They stick to tasks and see them through to completion, regardless of difficulties. They are always optimistic and confident that they can find solutions to problems. They may even be a little bit stubborn when they are convinced of the correctness of a decision.

Sense of responsibility: Leaders place responsibility above personal desires.

Emotional stability: Leaders must exercise self-control if they expect to control others, and must maintain control in the most trying situations. They should never allow personal problems to color decisions.

Enthusiasm: A leader must be genuinely enthusiastic in all the tasks which comprise the mission at hand. Followers will automatically give of themselves and take pride in their work when they know their leader is involved and committed.

Image: A leader must have a positive self-image.

Ethics: Ethics play a key role in leadership because they are the basis of all group interaction and decision making. Professional ethics require leaders to maintain high standards of personal conduct and to adhere to those standards in all situations so that followers can rely on the leader's actions. Leaders should not use their position for personal and special privileges.

Recognition: Leaders recognize the accomplishments of their people. William James, the philosopher, said, "The deepest principle in human nature is to be appreciated." Good leaders are aware of the people surrounding them. They know the names of subordinates, their home towns, family situations, and so on. They are aware of the feelings of their subordinates.

Sensitivity: Leaders must be aware of their own psychological and physiological states and be sensitive to the impact they have on others. They should be particularly sensitive to departures from their norms brought on by stress or fatigue. They must also be sensitive to the psychological and physiological states of others, and be prepared to adjust their style accordingly.

Flexibility: A leader must understand that no two people or two situations are ever exactly alike. Yesterday's solution may or may not be the correct solution for today or tomorrow. Effective leaders adapt their styles to the particular person or problem at hand.

Humor: Leaders should have a sense of humor, because they set the tone for their aircraft. Humor can be a positive and welcome contribution to an efficient and effective cockpit.

Stamina: Leaders have a high level of physical and mental stamina. Good leaders always seem at the ready and require only normal periods of rest. They know how to pace themselves well and maintain themselves in good physical condition.

This list may seem daunting, but it is intended to serve as a guide. Each of the concepts can be learned and, as you develop each, your ability as a leader will improve.

THE TEAM CONCEPT

When there are two or more crew members in a cockpit you have a team that must work together to do the best job possible for the team rather than working separately as individuals. This is as true for the single-engine airplane with a pilot friend in the right seat as it is for transport category aircraft. The primary reason for using a team as opposed to an individual to crew an airplane is to share the work, so that everything gets done in a more timely and effective way.

Learning teamwork is often difficult for pilots because virtually all of their experience is as an individual. All early training is based on a single-crew concept, and most evaluations (flight checks) are made on an individual basis. This strong emphasis on individual performance may, at times, be detrimental to effective team effort, because there has been no experience of sharing responsibility for making decisions and taking actions. However, when accidents occur, the crew is evaluated as a team — they go down together, most likely with the same fate. Sadly, this is the only consistent team performance evaluation being used today.

In the general aviation arena, some *ab initio* pilot training programs (those that teach airline procedures from the beginning of training) are breaking out of the traditional, single-pilot concept while operating within the requirements of current regulations. From the very beginning of their flying careers, students fly as part of a team, initially with an instructor in the right seat and a second student in the back seat. The second student is not just an observer, but an active participant in the flight. The second student participates in challenge-and-response checklists and is required to provide a critique of the pilot's performance using a set of printed performance criteria, such as heading, altitude, angle of bank, power setting, and so on. By encouraging teamwork and constructive criticism from the beginning, students build a habit of cooperation rather than competition.

One of the most important functions of an effective leader is to develop a team concept within his or her group. By team concept we mean a feeling or motivation to accomplish team goals over and above individual goals.

An excellent exercise for learning the team concept is to attempt synergism in the performance of a task. Synergism refers to a process by which the combined product of a group of people is greater than any individual effort. Most CRM courses have exercises, such as "The Caribbean Island Survival Exercise," that can be used for this purpose. These types of exercises consist of a problem, from either inside or outside of aviation, that each member is asked to perform individually. Then, the small group or team is brought together to attempt the solution as a team. If the team scores higher than the best individual within that team, it has achieved synergism. Synergism can best be achieved when all five of the communication concepts mentioned above (inquiry, advocacy, listening, conflict resolution, and critique) have been used by all individuals within the team.

DELEGATION

Another essential aspect of a good cockpit leader is an ability to delegate tasks to others, both for division of labor (task orientation) and to provide variety and experience to the other members of the crew (relationship orientation). In the cockpit, the captain has the primary responsibility for delegation. Although he or she gives up the specific task to other members of the crew, he or she still retains responsibility for its completion.

Delegation must be carefully planned and the plan must be flexible to accommodate changes that take place throughout the mission. In planning, the delegator should determine and communicate with other crew members:

* What has to be done.
* Why each task has to be done.
* The standard that the delegator requires.
* When each person is to perform the delegated tasks.
* The priority of each of the delegated tasks.
* What resources are available to complete each task.
* Who is responsible for doing each task.

Delegation should be made to each person according to interests, capabilities, and qualifications, and policy requirements — not simply according to position of authority in the cockpit.

As the mission progresses the delegator should not interfere with subordinates, but should monitor their performance in a non-threatening way and make adjustments to assignments as necessary. If mistakes are made, these should be corrected constructively.

THEORY OF THE SITUATION

Central to many CRM accidents is a discrepancy between what the crew perceives to be true about a given situation and what in fact is reality. Furthermore, as we indicated earlier in this chapter, the situation often determines the type of communication style you should use. To help you recognize these situations, a concept called "Theory of the Situation" is offered. This concept, developed by Dr. Lee Bolman of the Harvard University School of Education (1979), is an attempt to show how the various forces interact as you attempt to gain an awareness of the situation.

One of the reasons for discrepancies between perceptions and reality is that a major function of the human perceptual system is to reduce and order the vast amount of information coming in through the senses, so that you can understand and respond appropriately. Unfortunately, this information-reduction and ordering process is not perfect and sometimes leads to mistakes or discrepancies from reality. The process is developed through many years of experience and is, therefore, different for every person. Perceptions of visual information are fairly consistent with all individuals. However, perceptions of situations based on cognitive material obtained through all the senses is not very consistent.

The following definitions are offered for the elements of this model:

> **Theory of the situation**: What you assume to be true, based on your perception of the facts you have at any point in time.

> **Reality**: The situation as it is in reality — often not fully known until after an incident.

> **Theory in use**: Your predictable behavior in a given situation that has been developed since birth.

> **Espoused theory**: Your account or explanation of your behavior.

Theory in practice: The set of skills, knowledge, and experience you call upon according to your theory of the situation.

Figure 9-3 shows the relationships among the elements of the model. If the theory of the situation is in line with reality, the flight crew's assumptions are correct, and safe flight decisions are likely. However, if the crew's assumptions are not correct, a discrepancy exists between their theory of the situation and reality, as indicated at the top of figure 9-3. The greater this discrepancy, the greater the danger. Most CRM accidents are a result of these discrepancies going unrecognized and magnified under stress.

The remedy for the discrepancy between the theory of the situation and reality is through testing assumptions. This means actively checking your understanding of the situation with other crew members, ATC, the instruments, and/or computers on board the aircraft.

A second discrepancy can exist between your espoused theory and your theory in use, which can also lead to decisional errors. For example, if you say you will do something and have no intention of doing it, you create great misunderstandings within the crew and great conflicts within yourself. The key to reducing this discrepancy is found in courses that help to identify the extent of this discrepancy and show you how to deal with it. Always say what you intend to do and do what you say you will do in the cockpit.

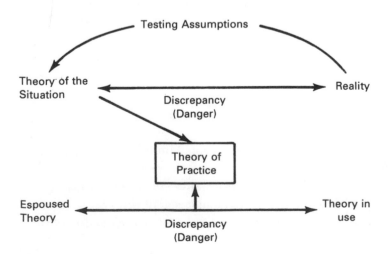

Figure 9-3

CONCLUSION

Dr. Robert Helmreich and his colleagues have given the attitude questionnaire at the end of this chapter to thousands of airline pilots and has observed hundreds of pilots perform their duties on the flight deck. As a result of these observations, he has made the following, somewhat surprising conclusions concerning the traits of an effective pilot-in-command and those of the ineffective pilot-in-command.

The effective pilot-in-command is one who:

* Recognizes personal limitations.

* Recognizes his or her diminished decision making capabilities in emergencies.

* Encourages other crew members to question decisions and actions.

* Is sensitive to personal problems of other crew members that might affect operations.

* Feels obligated to discuss personal limitations.

* Recognizes the need for the pilot who is flying the airplane to verbalize plans.

* Recognizes the captain's role in training other crew members.

* Recognizes the need for a relaxed and harmonious flight deck.

* Recognizes that the optimal management style varies as a function of the situation and fellow crew members.

* Stresses the captain's responsibility for coordinating cabin crew responsibilities.

The ineffective pilot-in-command is one who:

* Is the stereotype of the "macho pilot" with the "right stuff."

* Does not recognize personal limitations due to stress or emergencies.

* Does not utilize the resources of fellow crew members.

* Is less sensitive to problems and reactions of others.

* Tends to employ a consistent, authoritarian style of management.

* Has a flight deck that is more tense than that of a good pilot-in-command.
* Has a flight deck reflecting far less team coordination than that of the good pilot-in-command.

Cockpit Resource Management is recognized by the airlines and the military as being an important component of safe flight. Increasingly, companies are providing training in CRM to their pilots in the hope that they can reduce the number of accidents and incidents caused by a breakdown of communication and decision making capabilities.

One of the problems, however, is that it is difficult to change ingrained habits. A typical pilot flying for an airline has thousands of hours of experience in single-pilot airplanes in which he or she was solely responsible for making decisions and acting upon them. In most cases, this pilot has had little or no training in the elements of teamwork. It takes time and dedication to break such long-standing habits.

We believe that many of the elements of CRM are applicable to general aviation, even though the "crew" may be a student, instructor, pilot, or a friend. Communication and coordination are still important in these situations, so we encourage you to pay attention to the principles outlined in this chapter and to try them out in your own flight situations.

Exercises

Personal Characteristics Inventory

The following questionnaire focuses on **relationship** versus **task** orientation, which is important in multi-pilot crew decision-making. Answer the questions according to the instructions provided as honestly as you can. The scoring method and interpretation of the Personal Characteristics Inventory are provided at the end of this questionnaire.

Instructions: The items below inquire about what kind of a person you think you are. Each item consists of a characteristic, with the letters A through E. The letters form a scale between the two extremes of the characteristic — never and always. For example:

	NEVER			ALWAYS	
Humorous	A	B	C	D	E

Circle the letter which best describes where you think you fall on the scale. For example, if you think you have no humor at all, you would choose A. If you think you are quite humorous, you should choose D. If you are average, you might choose C, and so on. Be sure to answer every question.

	NEVER			ALWAYS	
1. Aggressive	A	B	C	D	E
2. Positive attitude	A	B	C	D	E
3. Independent	A	B	C	D	E
4. Arrogant	A	B	C	D	E
5. Emotional	A	B	C	D	E
6. Dominant	A	B	C	D	E
7. Modest	A	B	C	D	E
8. Excitable in a crisis	A	B	C	D	E
9. Active	A	B	C	D	E
10. Egotistical	A	B	C	D	E
11. Devoted to others	A	B	C	D	E
12. Timid	A	B	C	D	E
13. Gentle	A	B	C	D	E
14. Complaining	A	B	C	D	E
15. Helpful to others	A	B	C	D	E
16. Competitive	A	B	C	D	E

	NEVER			**ALWAYS**	
17. Commanding	A	B	C	D	E
18. Worldly	A	B	C	D	E
19. Generous	A	B	C	D	E
20. Humane	A	B	C	D	E
21. Need for approval	A	B	C	D	E
22. Democratic	A	B	C	D	E
23. Feelings easily hurt	A	B	C	D	E
24. Nagging	A	B	C	D	E
25. Empathetic	A	B	C	D	E
26. Decisive	A	B	C	D	E
27. Imperturbable	A	B	C	D	E
28. Tenacious	A	B	C	D	E
29. Optimistic	A	B	C	D	E
30. Cry easily	A	B	C	D	E
31. Self-confident	A	B	C	D	E
32. Self-centered	A	B	C	D	E
33. Feeling of superiority	A	B	C	D	E
34. Hostile	A	B	C	D	E
35. Understanding of others	A	B	C	D	E
36. Sociable	A	B	C	D	E
37. Dictatorial	A	B	C	D	E
38. Need for security	A	B	C	D	E
39. Gullible	A	B	C	D	E
40. Calm under pressure	A	B	C	D	E

SCORING THE PCI

To score your PCI, 16 questions are used. Eight questions are used to evaluate your Task orientation and eight are used to show your Relationship orientation. On the following PCI Scoring Sheet, write down your letter response to the question indicated. The letter response is then converted to a number using the following formula: A=0, B=1, C=2, D=3, E=4. Write your number response next to the letter response. Total your scores for Task and Relationship at the bottom.

PCI Scoring Sheet

TASK			RELATIONSHIP		
Question	Letter	Number	Question	Letter	Number
3			5		
9			11		
16			13		
26			15		
28			20		
31			25		
33			35		
40			36		
TOTAL			TOTAL		

INTERPRETING THE PCI

To determine where you stand on the following Task Versus Relationship Matrix, locate your Task score along the horizontal scale and your Relationship score along the vertical scale. Then, extend lines from these points into the matrix. The intersection of the two lines is where your personal characteristic is located. If your Task score is 21 or above, you would be considered high on task orientation. If your Relationship score is 23 or higher, you would be considered high on relationship orientation. Lower scores than these would be considered low on either scale.

Your score on this test should not be considered positive or negative as far as your capability as a pilot is concerned.

However, if you are very low on either scale, you might consider looking carefully at your thought patterns in the cockpit. In tests of professional pilots, about 90 percent score high on task orientation and about 50 percent score high on relationship orientation.

Task Verses Relationship Matrix

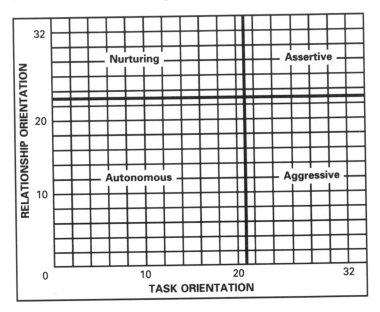

Helmreich Pilot Attitude
Questionnaire (Abbreviated Version)

1. My decision making ability is as good in emergencies as in routine flying situations (Agree-disagree).

2. Captains should encourage their First Officers to question procedures during normal flight operations and in emergencies (Agree-disagree).

3. Pilots should be aware of and sensitive to the personal problems of fellow crew members (Agree-disagree).

4. The Captain should take control and fly the aircraft in emergency and non-standard situations (Agree-disagree).

5. There are no circumstances (except total incapacitation) where the First Officer should assume command of the aircraft (Agree-disagree).

6. First Officers should not question the decisions or actions of the Captain except when they threaten the safety of the flight (Agree-disagree).

7. The pilot flying the aircraft should verbalize his or her plans for maneuvers and should be sure that the information is understood and acknowledged by the other pilot (Agree-disagree).

8. Pilots should feel obligated to mention their own psychological stress or physical problems to other flight crew personnel before or during a flight (Agree-disagree).

9. Captains should employ the same style of management in all situations and with all crew members (Agree-disagree).

10. Conversation in the cockpit should be kept to a minimum except for necessary operational matters (Agree-disagree).

11. Instructions to other crew members should be general and non-specific so that each individual can practice self-management and can develop individual skills (Agree-disagree).

12. Training is one of the most important responsibilities of a Captain (Agree-disagree).

13. A relaxed attitude is essential to maintaining a cooperative and harmonious flight deck (Agree-disagree).

14. The Captain's responsibilities include coordination of cabin crew activities (Agree-disagree).

15. The Captain should provide clear, direct orders concerning procedures to be followed in all situations (Agree-disagree).

Chapter Questions

1.   What is the ultimate objective of CRM training?

2.   Compare and contrast how we deal with personality and attitudes in CRM training.

3.   Describe the four behavioral dimensions of the Task versus Relationship Model: Nurturing, Aggressive, Autonomous, and Assertive.

4.   What are the five important factors of good cockpit communication? Briefly describe each.

5.   What are the signs that someone is a poor listener? A good listener?

6.   What is synergy in CRM?

7.   What is the role of leadership in CRM?

8.   Research one of the accidents listed on page 9-2 in this chapter. Explain how poor CRM contributed to the accident. What could have been done by each crew member to help avoid the accident?

9.   What are the basic principles of conflict resolution?

10. Describe how you would provide a critique of a colleagues performance without creating hostility.

11. What are the qualities of a good leader?

12. As a leader, how would you foster teamwork?

REFERENCES AND RECOMMENDED READINGS

Helmreich, R.L., Foushee, H.C., Benson, R., and Russini, W. 1986. "Cockpit management attitudes: Exploring the attitude-behavior linkage." *Aviation, Space, and Environmental Medicine*, 57, 1198-1200.

Jensen, R.S. 1989. *Aeronautical decision making - Cockpit resource management.* Springfield, VA: National Technical Information Service, DOT/FAA/PM-86/46 Final Report.

Jensen, R.S. and C.S. Biegalski. 1989. "Cockpit resource management." In: R.S. Jensen (Ed) *Aviation Psychology*. Aldershot, UK: Gower Technical Press.

An excellent exercise for teaching synergism is:

The Caribbean Island Survival Exercise
Designs for Organizational Effectiveness
P.O. Box 1146
Fairfax, VA 22030
703-691-4056

An excellent test for leadership is:

The LEAD Test (Hersy and Blanchard)
Learning Resources Corporation
8517 Production Ave.
P.O. Box 26240
San Diego, CA 92126
714-578-5900

The Future 10

INTRODUCTION

This chapter is different from the others in the book because we look into the future at issues that relate to the topics we have discussed in this book. We start by taking a brief look at pilot training and how it may evolve over the next ten years. We look at the future of human factors training, as well as some advances in technology that promise to improve the quality of training. We also take a close look at some of the underlying issues of cockpit design. This is particularly helpful as we try to understand the rapidly changing nature of aircraft design with respect to automation in both commercial and general aviation airplanes. Increasing automation has a dramatic impact on the role of the pilot and can change the whole nature of flying. As you start evaluating how automation affects you, such as when you are buying an airplane, it is helpful to have a good base of knowledge in these design principles.

TRAINING IN THE FUTURE

We see two major changes in how flying is taught in the future. First, we believe that there will be an increasing emphasis on human factors and topics this book discusses in classroom training for pilots. Second, technological training devices will become more prevalent, because they will incorporate powerful instructional ideas that make training more effective and efficient.

HUMAN FACTORS TRAINING

We mentioned earlier that the International Civil Aviation Organization (ICAO) has called for the inclusion of human factors as a formal part of pilot training throughout the world. ICAO would like to see this happen for people who are learning to fly and as refresher training for current pilots. Human factors will have the same status as the traditional knowledge areas of meteorology, navigation, and principles of flight. Furthermore, ICAO recommends that licensing requirements include the evaluation of certain human factors skills, such as the demonstration of good judgment, the application of good crew coordination skills (where applicable), and the use of effective communication skills.

Figure 10-1 contains the outline for the ICAO syllabus for ATP certification. As you can see, it is a comprehensive course lasting 35 hours. For other certificates, such as the Private or Commercial, the syllabus can be modified to accommodate the appropriate requirements and the time can be reduced.

We are excited about this trend toward formal ICAO recognition of human factors as a major topic in aviation training. We are convinced that it will have a demonstrably positive impact on aviation safety.

TECHNOLOGY AND TRAINING

Earlier, we discussed computer-based training (CBT) as one of the resources you should look for as you evaluate flying schools. We said that two methodologies were common in the use of CBT in ground schools: namely, tutorial instruction and computer-based testing. Both of these techniques

Module 1:	**Introduction**	1.75 hours
Module 2:	**Man (Physiology)**	7.00 hours
Module 3:	**Man (Psychology)**	10.50 hours
Module 4:	**Man (Fitness)**	1.75 hours
Module 5:	**Pilot (Equipment)**	1.75 hours
Module 6:	**Pilot (Software)**	3.50 hours
Module 7:	**Interpersonal Relations**	5.25 hours
Module 8:	**Operating Environment**	3.50 hours

TOTAL: 35 hours

Figure 10-1

have been around since the 1960's and have been found to be effective if well designed and implemented. In the flight portion, computer-based simulations offer promise as a way to learn procedures and skills in a less expensive and nonthreatening environment.

COMPUTER-BASED TRAINING

The use of computers for learning is growing at an accelerated pace outside of aviation and is beginning to make changes in aviation as well. However, we urge you to exercise some caution, especially if you are buying computer programs to enhance your flight training and experience. Many instructional computer programs are not well designed and can be more frustrating than helpful. Before you spend your money, talk to people who have used the material and, if possible, review it yourself to ensure it is the quality you need.

One other potential drawback of instructional software is that it can reduce your interaction with your instructor. We think that you should treat the computer as a supplement to your traditional instruction, not a replacement for it. The industry has not yet achieved the sophistication of instructional design for computers to replace teachers.

Notwithstanding these cautionary comments, we are strong advocates of computers in training and believe that they can enhance the quality of training. For one thing, they allow you to learn material at your convenience at a pace that suits your learning style. They are also infinitely patient, which is helpful for many people.

There are some areas where computers will become increasingly helpful in aviation training, one of which is human factors. In the future, you will be able to go through programs designed to teach you good judgment. We can conceive of students being presented lifelike scenarios that require a series of decisions, just as happens on a normal flight. As you make these decisions (or do not make them), the scenario changes to reflect your input. The scenario can also be manipulated by the computer to reflect changes in the underlying status of the flight, such as changes in weather patterns or airplane functioning.

In programs like these, you would continue until your flight terminated at its logical conclusion, whether it was successful or not. The computer could then go through each of your decisions, helping you understand the implications of your choices and pointing out what other options you had.

An example of this type of program deals with preflight planning (Gibbons, Trollip, & Karim, 1990). At the start of the program, you are given a particular mission to fly, such as a VFR flight from one city to another. You are told the type of airplane, the number of passengers you have and their weights, and the weight of the baggage. Your goal in the program is to file an appropriate and accurate flight plan for the trip.

The program then pictorially presents you with four options: you can visit the flight service station (FSS), you can go to the airplane, you can access materials in your flight bag, or you can complete worksheets concerning the flight. At any stage, you can file the flight plan.

If you choose to visit the FSS, for example, you can request current and forecast weather information, including teletype printouts and weather maps, as well as NOTAMs. You can read relevant chapters of the *Airman's Information Manual* and listen to ATIS. If you choose to go to your airplane, you can access maintenance records, licensing and registration documents, any placards, and the current tachometer reading.

In your flight bag, you have your own pilot and medical certificates, the airplane operating manual, sectional charts, and so on.

The worksheets allow you to fill out the necessary information and make the calculations for center of gravity, time enroute, and fuel consumption. They also give you the opportunity to use graphics to plot your intended course with waypoints. You can also tell the computer your intended altitudes for each segment of the flight.

Finally, you can file your flight plan using the standard format. To this point, the computer has done nothing but provide you access to the information you would normally have when planning a flight. At no time did it provide feedback as to the appropriateness of any of your responses.

When you file the flight plan, all the information that the computer has been gathering, including which information you requested, which calculations you did on the worksheets, and so on, are passed to another program for evaluation. First, this evaluator program (which may be considered an "intelligent," computerized instructor) checks all the current information (the same that you had access to) and decides whether it would have filed a flight plan and, if so, whether it would have filed the same type of flight plan (IFR, VFR, routing, altitude, etc.). It then compares yours with its own. If the two match, you have performed well. If not, it compares each step that you took against an ideal sequence.

Whenever there is a mismatch, the program tries to find out from you how you came to a particular decision. For example, if your time enroute was different from the one the computer calculated, the program would ask you for the specific information you used in your calculation. It would ask you to give the distance of the flight, the cruise airspeed, the time for climbing and descending, and the fuel consumption of the airplane. Wherever possible, it would try to lead you to finding your own mistakes, letting you correct them rather than telling you where you went wrong. If necessary, it could provide remedial instruction on an area where you were having trouble.

In addition, if you filed an accurate flight plan, but had failed to check the maintenance records to ensure that the airplane was airworthy or not in need of maintenance, the program would remind you of this lapse of judgment. Obviously, it would do the same if you neglected to check the weather or loaded the airplane so that the center of gravity constraints were violated.

A program like this allows you to go through your normal preflight planning procedures, in a nonthreatening and instructive way. By having an "intelligent" instructor built into the program, it is able to make assessments about your ability to make good decisions and exercise good judgment.

Computer-based training is only in its infancy in aviation, but it has the potential of having a significant impact on areas that are procedural or attitudinal in nature. We expect to see big strides in the use of this technology in the near future.

SIMULATION

Another area in which computer technology is beginning to play a big role in improving our capabilities for flight training is simulation. Most simulators today are based on computer technology, as are their visual displays. However, computers can provide added capabilities that will improve the efficiency and effectiveness of training.

Earlier, we discussed the fact that a good visual display adds to the benefits of a simulator. However, properly configured it can do so much more than present a view of the outside world.

Some newer visual systems incorporate factors specifically designed to facilitate learning. Dr. Gavan Lintern, at the Aviation Research Laboratory at the University of Illinois, for example, has developed a display specifically designed to help a student learn to land. The computer program that drives the display also monitors the position of the landing airplane with respect to an ideal approach. When the student's approach becomes too far off, the computer displays a set of guide posts on the screen to indicate visually and nonintrusively to the student where the airplane

Figure 10-2

is and where it should be. [Figure 10-2] When the student makes a correction and the airplane moves back onto an acceptable approach, the guide posts disappear again. Lintern has shown that this system can teach student pilots so effectively that many are able to land the real airplane successfully in the first or second lesson.

This idea of building instructional benefits into traditional training technologies has a good future in the aviation industry. For example, the Hawker Pilot Trainer, a general aviation simulator first developed in Australia by Mr. Jim Sparks of Aircraft Training Systems and Dr. Stan Roscoe of Illiana Aviation Sciences and manufactured by Hawker de Haviland, not only incorporates Lintern's augmented landing displays, but also has almost all the instruments displayed on large CRT's. [Figure 10-3] This gives the instructor the

Photo courtesy of Hawker deHaviland, Australia

Figure 10-3

flexibility to present information that is helpful instructionally because the displays can be modified relatively easy through changes in the software. For example, if the instructor would like to show the effect on the lift-to-drag ratio of changes to pitch and roll, he or she can select that mode and the computer will display a lift-to-drag curve that moves in response to control movement. It literally makes aerodynamics come to life!

In the future, different instructional scenarios will be readily accessible from a library in the simulator's computer, enabling students to select the type of environment in which they want to fly. An added benefit of having these instructional features designed into the simulator is that a student can use the simulator without an instructor and still receive useful feedback.

Aviation has always had an anomalous attitude toward instruction. On the one hand, the industry is always on the cutting edge of technological innovation, both for airplanes and training devices such as simulators. On the other hand, it has always been extremely conservative in changing the instructional methods used. For the most part, instructional techniques are no different today from what they were 40 years ago. Computer technology is beginning to change this, because it is able to incorporate sound instructional methods without the problem of having to retrain instructors. We expect to see dramatic improvements in future training due to increased use of devices like the ones described above.

COCKPIT DESIGN

Throughout the book, we have made the case that many errors, classified as pilot errors, were the result of poor design. In the chapter on Cockpit Design, we introduced this idea with a number of examples. In this section, we elaborate on these ideas by introducing some human factors principles related to cockpit design.

You may wonder why we think such a discussion is important in a book for general aviation pilots. There are several reasons. First, the more you understand about how instruments are designed, the better equipped you will be to evaluate new designs that come on the market. Second, it is quite possible that some time in the future you will be asked to make suggestions for new instruments or for new designs of instruments. You may have seen the advertisements in many popular aviation magazines showing a picture of an instrument panel with the caption, "Designed by pilots!" Many aircraft manufacturers use pilot inputs into their cockpit designs. An understanding of design principles will make your suggestions more valuable and valid and, perhaps, it will make you willing to step beyond the traditional cockpit designs that have had little human factors input. Finally, we think that having a better grasp of design will improve the way you interpret current instruments in your everyday flying.

In this book, we want to give you a sense of the human factors issues surrounding design. To do this, we will discuss one aspect of design that is becoming increasingly prevalent: namely, automation.

HUMAN CAPABILITIES

The first step in understanding design based on human factors principles is to understand something about a human's capabilities — to have a good grasp of what tasks people are good at and tasks for which they are not well suited. Figure 10-4 shows the four phases of the information processing model introduced in Chapter 1, figure 1-3. In this section, we discuss the tasks that typically occur in each phase. This information is invaluable to designers because, it enables them to make the machine fit the operator, helping where capabilities are weak and being responsive in areas in which capabilities are strong.

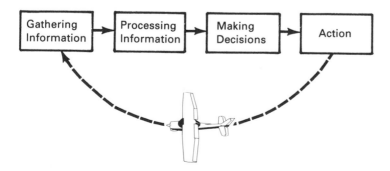

Figure 10-4

Gathering Information

The pilot requires information about the current performance and status of the airplane to operate it effectively and safely. This information needs to be presented to the pilot in a form that is easily interpreted so that decisions about control movement can be made without question. The following table indicates well established principles about the relative capabilities of humans and machines to gather certain types of information.

Capability	Human	Machine
Detect very small changes in visual or auditory information	X	
Detect targets with a lot of background noise, such as aircraft lights over a city	X	
Detect very short or very long sound waves, such as x-rays and radio signals		X
Recognize small changes in complex patterns, such as pictures or sounds	X	
Monitor for predetermined event, such as higher-than-normal fuel consumption		X
Sense unusual or unexpected events	X	

As you can see from the table, humans are generally better at interpreting visual and aural information than machines, especially when the information is masked by noise or when changes are small. On the other hand, machines are better at dealing with detecting wavelengths outside the human range, which is not surprising. From a design perspective, this means that tasks, such as avoiding other aircraft, should be the responsibility of the pilot, unless all aircraft have electronic means of detecting each other.

As it stands, the current system for traffic separation follows these design guidelines. For example, in the cockpit, the pilot has to maintain constant vigilance for other aircraft, and on the ground, the air traffic controller detects targets on the radar screen (with computer assistance). When electronic proximity alerting systems, such as TCAS, are implemented, computers will play an increasingly important role in collision avoidance.

A second part of information gathering is the ability to detect changes in the status of predetermined events, such as higher-than-normal fuel consumption during a flight. As you can see from the previous table, humans are not good at doing this, particularly when they have to spend long hours watching for very small changes. Human attention is limited and boredom sets in if the task is tedious. On the other hand, machines are very good monitors of performance and can pay undivided attention indefinitely.

Finally, humans are better at noticing unexpected or unusual events. This is one of the compelling reasons for keeping pilots in the cockpit. Computers can fly the airplane more accurately and more efficiently than humans but, should something happen that is unanticipated, it is likely that a human will be able to resolve the problem more easily than the computer.

These last two points raise a fundamental issue of how airplanes are designed. The current trend, especially for large, complex aircraft, is for most of the flying to be handled by the on-board computers and autopilots. This leaves the pilot to monitor whether all the systems are functioning properly. This contradicts what we know about human performance. People are not good at extended periods of monitoring, while computers are. Furthermore, even if a human notices a problem after hours of monitoring, it is difficult for that human to take over control, because he or she is unlikely to have a feel for the status of all of the systems. Our position is that it would be a better design for the human to fly the airplane routinely and have the computer monitor the performance. In this way, the pilot is always in the loop and knows exactly what the status is of all the systems. If attention and performance lags, the computer can promptly bring this to the attention of the pilot, who can then make the necessary corrections.

Processing Information

After a pilot has gathered information, he or she needs to process that information. The following table lists the factors relevant to information processing and indicates the relative capabilities of humans and machines at performing the particular task.

Capability	Human	Machine
Store generalized information over a long period, such as principles or strategies	X	
Store detailed information over a long period, such as V-speeds or performance data		X
Reason inductively	X	
Make subjective estimates or evaluations	X	
Prioritize tasks in times of overload	X	
Reason deductively		X
Retrieve stored information quickly and accurately		X
Perform large computations quickly		X

As you can see from this table, the major difference between the processing capabilities of humans and machines lies in the level of sophistication. Machines are better at remembering specific information and retrieving it quickly, as well as at doing computations and using established rules to generate solutions. Humans, on the other hand, are good at almost exactly the opposite things: they remember principles better than facts; they remember things slowly and often inaccurately; they are not good at computing, but are good at making inferences; and they can exercise judgment as to how they allocate their attention to different tasks.

From a design perspective, this means that pilots should not have to remember detailed information accurately; in fact, that is why checklists are so valuable. It is also why you should write down information given by ATC. Such information often comes quickly and in amounts too large to remember accurately.

Making Decisions

The third part of the information processing model is making decisions. In many ways, this is similar to the processing stage. However, issues of judgment and choice also play a role. As you can see from the table below, this is an area in which humans can outperform machines.

Capability	Human	Machine
Reason inductively	X	
Make subjective estimates or evaluations	X	
Prioritize tasks in times of overload	X	
Devise strategies to solve new problems	X	

The implication of this table is that humans should be given the role of making decisions when uncertainty is present. Machines make good decisions when the rules and conditions are known, but they do not perform as well as humans in times of uncertainty. Furthermore, humans are able to place values (often a subjective issue) on different decisions and they use these values to make the final choice. Machines cannot do this.

Action

The final phase of the model is taking action on the decisions. The following table lists various actions and indicates the relative abilities of humans and machines to carry them out.

Capability	Human	Machine
Make rapid and consistent responses to given signals		X
Perform repetitive activities reliably		X
Maintain performance over long periods of time		X
Perform several activities simultaneously		X
Maintain efficient operation under conditions of heavy load		X
Maintain efficient operation under conditions of distraction		X

This table indicates that machines almost always outperform humans when it comes to an operational environment. And this is where one of the fundamental conflicts arises from a design perspective. If machines (computers) can fly an airplane better than a human, but humans need to be in the loop in order to recover from a situation which the machine cannot handle, how can the designer provide for both?

One answer discussed later is "semi-automatic," as shown in figure 10-8. What most transport category aircraft designers have opted for is to leave the pilot out of the loop and have the computer perform most of the control tasks. In part, this decision is based on the fact that the computer can fly more economically, as well as more smoothly, than a human pilot. The potential danger of this decision is that in times of equipment malfunction, the pilot may not be able to assess all relevant facts quickly enough to take over and correct the situation.

Throughout the rest of the chapter, keep in mind or refer back to these lists of capabilities as you read about the advantages and disadvantages of automation. Try to draw your own conclusions about whether certain design attributes maximize the capabilities of the human operator and the machine — in our case, the pilot and the airplane.

AUTOMATION

In the following pages, we discuss the intriguing topic of automation as it relates to issues of design and human performance. We hope that the discussion leads you to a better understanding of some of the fundamental issues facing aircraft designers. Some of the ideas expressed are those of Dr. Earl Wiener, one of the foremost experts on automation in the cockpit today (Wiener and Nagel, 1988).

As pressure for automation increases, it will become increasingly important for designers to adhere to principles designed to minimize problems for the human operator. The application of automation, as in any technological addition to the cockpit, requires a careful examination of the impact it may have on the ability of pilots to perform the tasks they have been assigned to do. From your perspective, you will face a dramatically changing cockpit over the next 10 to 20 years. The following automation principles should help you evaluate how well the new systems have been designed to suit your needs. Taken as a whole, these principles have become known as "Human Centered Automation."

DEFINITION

There many definitions of automation. The most obvious one is the replacement of manual tasks by machine-controlled functions.

A common example of automation in light airplanes is autopilots. These can range from simple wing-levelers that take the place of manual roll control during straight flight, to autopilot systems that fly automatic approaches, replacing manual flight control all the way down to minimums on an approach. In transport aircraft, modern autopilot systems have the capability of flying the airplane from lift-off to touchdown at a distant airport without the pilot ever touching the flight controls. They replace all manual control except voice communications and an occasional change in the computer program as the flight proceeds.

A second definition of automation refers to the presentation of cockpit data on cathode ray tubes instead of mechanical or electromechanical instruments. These so-called glass cockpits are being installed into virtually all airliners and many high-end, general aviation airplanes. As you can see in the cockpit shown in figure 10-5, virtually all flight, navigation, and systems information is presented on computer screens. Mechanical flight instruments are present for use in the event of a total electrical failure.

A third definition of automation refers to computational support for flight activities, offered by on-board computers. The first application of computers to flight was through the introduction of area navigation, which allowed the creation of "phantom" waypoints and enabled you to fly direct routes to your destination. Today, many general aviation airplanes have LORAN C units that provide the same capabilities. These units have relieved pilots of much of the uncertainty and stress of navigating in unfamiliar areas.

Photo courtesy of Collins Avionics, Rockwell International Corp.

Figure 10-5

In larger airplanes, computers are also being used to determine and fly the most efficient climb and descent gradients providing greatly improved fuel efficiency.

A fourth definition of automation refers to cockpit monitoring and alerting systems. There are numerous alerting systems on board both general aviation and airline aircraft. These are sometimes referred to as "bells and whistles." Most of these systems are aural, some are visual, and some are both aural and visual. Their purpose is to alert the pilot to the passage of certain planned positions or altitudes and of impending dangerous situations.

All airplanes have stall warning devices to alert the pilot that the angle of attack is approaching a stall, and emergency locator transmitters that emit a radio signal when they are set off by an impact. Retractable gear airplanes have gear-up warning devices to alert the pilot that the gear is up when the throttle is reduced below a certain power level. And there are alerting systems that tell the pilot when a preset altitude or waypoint is being approached.

Large airplanes, especially the newer glass-cockpit airplanes, have many other alerting systems including ground proximity warning systems (GPWS), flap position warnings, master cautions, and so on.

LEVELS OF AUTOMATION

In some of today's general aviation airplanes and in many transport category aircraft, there are three levels of automation available to the crew. Each has certain advantages and disadvantages, depending on the situation of flight. Figure 10-6 shows a block diagram of the configuration of the crew/aircraft control system in the **manual mode**. This diagram is just like the one shown as a general man-machine system diagram in Chapter 3, figure 3-1. Going around the loop, the crew adjusts the cockpit controls (aileron, rudder, elevator, flaps, etc.), which send signals to the control system actuators, which move the control surfaces, which cause changes to the aircraft attitude, which are detected by sensor systems (pitot-static, gyros, etc.) and sent to the display systems (including the view out the windscreen), and then back to the pilot through his or her own sensing system.

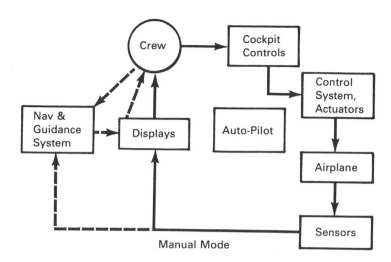

Figure 10-6

In the manual mode, workload tends to be high during certain portions of the flight, especially during the approach and landing phases.

For comparison, figure 10-7 shows the same set of components in the **fully automatic mode**. In this case, the cockpit controls are outside of the control loop and are not used. Instead, the autopilot system controls the aircraft with programming inputs from the crew through the navigation and autopilot systems. Workload is low (perhaps too low for optimum performance), and the crew is largely out of the control loop.

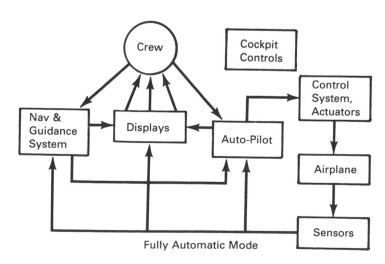

Figure 10-7

Finally, a **semi-automatic mode** is shown [Figure 10-8] because it is being used as an alternative to both the other options. It offers some real advantages from the standpoint of human factors. In this configuration, the crew stays in the control loop through the autopilot. A system called "control wheel steering" (CWS) controls the most difficult aspects of the task and allows the pilot to control such factors as the rates of turn, climb, and descent.

Automation is clearly the direction the aviation industry is taking. However, there are serious questions being raised by human factors specialists concerning the impact of automation on pilots and their role in the cockpit. The following discussion offers some insights into the advantages and disadvantages of automation in aviation.

ADVANTAGES OF AUTOMATION

Automation has many advantages in the cockpit. It can perform many of the continuous control tasks, freeing the pilot for more important functions, such as decision making. It can also be a boon to safety by removing human error at the source, replacing fallible humans with virtually unerring machines — a compelling argument, since pilot error is the reason for such a high percentage of accidents.

Other advantages of automation in the cockpit include reducing the crew size on transport aircraft, reducing the size of the cockpit, and providing better fuel management, all of

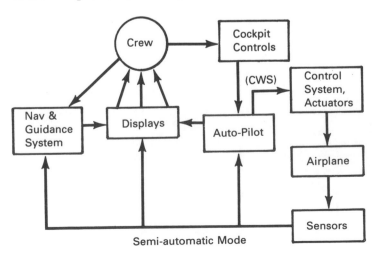

Figure 10-8   Semi-automatic Mode

which help to reduce costs. Automation also results in smoother, more accurate control of the airplane than can be achieved by a human. In addition, automation provides more options for presenting information through computer-controlled and programmable displays.

The industry has weighed these advantages with the disadvantages, and has concluded that automation is the course to follow. We now examine the disadvantages associated with automation.

DISADVANTAGES OF AUTOMATION

The almost unanimous response from experienced pilots, when asked about automation, is that it is reducing the pilot to the status of a button pusher and is stripping the job of its meaning and satisfaction. It is certainly important to take pilot satisfaction into account because low satisfaction can lead to poor performance. However, our primary interest is to discuss automation in light of its potential for impinging upon safety.

A common argument against the introduction of automation is that the replacement of an intelligent human by computers programmed by engineers can never be as safe as the situation when the human has full control. Proponents of this position argue that there is a growing list of accidents in the airline industry that can be attributed to automation. Accidents such as:

1.   Eastern Airlines DC-10, 1972, in the Everglades, where the autopilot inadvertently disconnected.

2.   Air New Zealand DC-10, 1981, into Mt. Erebus, where incorrect navigation data were entered into the computer by ground personnel.

3.   China Airlines B-747, 1985, over the Pacific Ocean, where there was a slow power loss in the number four engine. The autopilot tried to hold heading and altitude, but eventually caused a stall and spin.

4. Northwest MD-80, 1988, in Detroit, where automatic systems were implicated in attitude commands during its attempt to stay airborne.

5. US Air 737-400, 1989, at New York LaGuardia where, on auto takeoff, the trim system was not set properly.

These accidents are examples of the problems caused by automation in the cockpit. In each of them, there are some important human factors reasons for the problems occurring.

Mental Workload Increase

Automation leads to a definite decrease in physical workload. However, it can also lead to an increase in mental workload. When monitoring an airplane flying itself, pilots must constantly try to figure out what the computers are doing. In addition, automation requires pilots to manually enter a great deal of information into the flight computers. This, too, can increase the mental workload by increasing the need to constantly check that these data are accurate.

With the increased number of procedures requiring the pilot to insert, check, and crosscheck information in the computer, there is an increased potential for error.

Manual Takeover Problem

Whenever the pilot is not in the loop, he or she is often not mentally prepared to take over the controls and fly the airplane in the event of automatic system failure. This problem is regarded as a possible factor in many of the accidents listed above. The China Airlines accident is a good example. In that accident, the crew was absorbed in other activities and did not notice that the autopilot was banking the airplane to try to correct for the differential thrust caused by the loss of power in one engine. The crew did not attempt to correct the problem until their Boeing 747 had entered a spin toward the ocean below. Fortunately, the crew managed to recover before hitting the water.

A second problem with not being in the loop is that the pilot does not receive the benefit of feedback through the control system concerning the status of various systems. A subtle change in control dynamics (such as, reduced engine power or a shift in the cargo) may not be noticeable if the airplane is on autopilot. In addition, under autopilot control, there is a tendency for the pilot not to crosscheck the flight instruments as frequently, because it is assumed that the autopilot is doing its job.

Skill Degradation

There is a definite tendency for the pilot to lose manual flying skills when most portions of every flight are made on autopilot. Some companies, in fact, encourage auto-coupled approaches because they are more comfortable to the passengers. In Europe, some airlines encourage the use of auto-land.

Complacency

Pilots must always guard against complacency. When activities become routine — and we certainly hope that most flights are routine — it is easy for people to relax a little and not put as much effort into their performance. Automation has the potential of lulling pilots into the expectation that all flights will be routine, because so much of the work is done by the systems. Pilots must be particularly careful to avoid this potential cause of error.

Boredom

Boredom is a dull state of mind that comes over a person when an unstimulating activity or environment is repeated over and over again for a long period of time. Boredom can lead to stress, fatigue, and a motivation to seek more stimulating activities. It can also lead to reduced job satisfaction and a lower self-concept. The danger in flight is that the bored pilot may find it difficult to stay awake, may lose his or her capability to be vigilant, and may lower his or her standards of performance. Boredom is a normal human reaction to unstimulating circumstances.

Automation in airplanes tends to take away stimulating tasks and replace them with tasks that are not very stimulating. In automated cockpits, therefore, boredom is a likely outcome.

False Alarms

Any warning system can and will make errors. If they are set too sensitively, they will yield false alarms; that is, they will send an alarm signal when nothing is wrong. If they are set too tolerantly, they will miss critical events and fail to do their job. Both events can reduce their usefulness in flight and can even lead to accidents.

The most notable example is the ground proximity warning system (GPWS) which has a signal that goes off saying "PULL UP! PULL UP!" when the aircraft is approaching the ground and it is not in the landing configuration. If this system is not properly tuned, it can yield false alarms that will cause pilots to become so irritated that they will turn it off to avoid the distractions and uncertainty.

In the discussion above, we have highlighted some of the advantages and potential disadvantages of automation in the cockpit. As you can see, the design issues surrounding the topic are difficult, because it seems that each advantage gained results in one or more potential problems. In the following section, we lay out some human factors principles that we feel should guide designers in making choices between the various options.

HUMAN FACTORS PRINCIPLES IN COCKPIT DESIGN

In addition to the automation principles given above, there are some other cockpit design principles that you should look for in your cockpit or suggest if you are asked to assist in cockpit design. Some of these principles are well grounded with strong research, others are given because we feel that this is where the future is leading us in cockpit design. If you do not see the following design principles used in your cockpit, you should know that special care is needed on your part to avoid making errors.

PICTORIAL REALISM

One of the most important design principles for the cockpit is known as pictorial realism. According to this principle, to reduce the transformations that our perceptual system must apply, instruments or displays should look as close to the real world as possible. This means that whenever possible, we should use concrete pictures and symbols rather than abstract figures. The more abstract the presentation, the greater the amount of mental transformation required.

The reason for this principle is that we have a high degree of visual orientation. We convert most abstract words, numbers, and symbols to graphical images before we make decisions about what action to take. If the information needed to make decisions were presented graphically in the first place, we would not need to make the mental transformation before making the decision. The transformation takes time and can lead to error.

Consider an extreme example. As you know, numbers, letters, and words are forms of abstract representations of our thoughts that we use to communicate with others. Suppose that the bank angle was presented in digital form in degrees of bank. Let's say left bank was positive, right bank was negative, and zero bank was zero as presented on a CRT in the form of a number in front of the pilot. Could you still control the airplane on instruments? The answer is that you probably could but with great difficulty, even with a lot of experience. Fortunately, we have a more concrete display in the form of an attitude instrument that allows us to convert the indications into control decisions more directly.

In a study of pictorial display principles Jensen (1981) presented on a CRT the information normally shown on several separate instruments (attitude indicator, turn coordinator, altimeter, and course deviation indicator). Using a pictorial format, he found that nonpilots could learn to manually fly a complex, curved approach with a severe wind shear in 30 minutes of training almost as well as a professional pilot!

This is not to say that abstract numbers and words should never be used. Status indications that do not change rapidly and other forms of communication can be more easily transferred through abstract symbols.

Furthermore, the principle of pictorial realism does not apply universally. One must also consider the message that is to be conveyed and gear the form to the message to make it meaningful to you, the pilot. Often, a "realistic" picture presents far more information than is needed and it may be totally disorganized. The map of a subway system, for example, does not need to be totally realistic. Instead, the information should be formatted according to what the

Figure 10-9

passenger needs to know. As shown in figure 10-9, the information departs from reality to avoid confusion but retains the realism of relative position of the items of importance.

Similarly, in the cockpit we should provide concrete representations that have elements of reality without unnecessary detail. Computers now have the capability of presenting a multitude of information in almost any form. Care must be taken to understand what the pilot needs to know and to present it in a way that leads to the quickest and most error-free decisions.

DESIGN FOR ERROR PREVENTION

Designers must recognize that humans can be expected to make errors in every operation they are asked to perform. The same error may only be made once by the same pilot, and not every error will be made by every pilot. But, when pilots are required to do operations, statistics show that some pilot some time will make an error in that operation.

The first principle, therefore, should be to design systems to reduce the chances for human error, and to make them less vulnerable to catastrophic failure when an error is made. This is called error-tolerant design. Mechanisms must be provided that check for possible errors and alert the pilot before they become catastrophic.

PROVIDE COCKPIT FLEXIBILITY

Automatic systems should be designed so they can be modified to suit the operating style of the individual pilot. Despite attempts at standardization, pilots have strong preferences on how they like to operate — preferences based on their training, habits, and personality. In some cockpits today and in the future, pilots will be able to change display symbology and other parameters of automatic systems to fit their own style.

PROVIDE GOAL SHARING

In order to monitor cognitive functions of the pilot, the computer needs to have in its memory the intentions of the pilot, such as destination, route, altitude, and so on. The computer can then compare these intentions with the behavior it sees the pilot exhibiting. It can also compare fuel, oil, and other expendables against intentions to determine whether or not the pilot is making the correct choices.

GRACEFUL DEGRADATION OF SYSTEMS

The principle of graceful degradation states that when systems fail, backup systems should be provided so the failure does not cause an immediate catastrophe. This means that, as a pilot, you can expect that when some system fails, you will not fall out of the sky. This is a special concern to pilots of the so called, "fly-by-wire" aircraft that are beginning to appear in the airlines and will eventually also be available to general aviation. Following this principle there should be three levels of degradation of system performance:

1. Fail/Operational — The first system failure is automatically switched to a backup system with no change in performance or task. The pilot is merely informed of the failure and change.

2. Fail/Operational — The second level of system failure requires manual switching, but permits the pilot to continue to operate with little change in the flying task.

3. Fail/Safe — The third level of failure requires a substantial shift to manual backup procedures, but can be performed safely by the pilot.

It is worth pointing out that the trend to automation is only just beginning. In the near future, techniques in the field of expert systems will find their way into the cockpits of both military and civilian airplanes. These computer systems will perhaps be the ultimate in automatic systems. In the military, computerized "pilot associates" are expected to replace the second pilot in fighter aircraft of the future. The second crew member in civil aviation may not last long after that.

We also want to point out that the above discussion only scratches the surface of issues related to human factors design. There are similar principles that relate to the design of displays, alarms, controls, and so on. All of these attempt to ensure that the system design does not introduce a new set of errors. Good design should promote improved efficiency and performance, as well as reduce the risk of error. However, this will only occur if the human factor is consciously taken into consideration.

CONCLUSION

In this chapter, we hoped to provide you with some brief insights into the directions aviation is heading. We strongly urge you to explore the issues more thoroughly in books devoted to each of the topics, as well as through discussions with industry leaders.

Flying is about people, much more so than about machines. All aspects of the aviation system are designed by people, and all components of it are operated by people. How people think and act, therefore, has an impact on the success of the system. In this book, we have introduced you to a variety of factors that can affect your performance as a pilot, sometimes for the better and sometimes for the worse.

We hope that we have aroused your interest in human performance, particularly your own, and encouraged you to find out as much as you can about the topics we discussed, as well as about others we could not include. We have no doubt that increasing your awareness of human strengths and frailties will make you a safer pilot.

━━━━━━━━━━━━━━━━━━━━━━━━━━━━━ **Exercises**

Chapter Questions

1. What are the major topics covered in the ICAO Human Factors syllabus?

2. Explain the benefits of Lintern's use of computers to teach landings.

3. List the types of activities that machines perform better than humans in gathering information.

4. Are current trends in airplane design compatible with the strengths and weaknesses of humans and machines? Support your answer with examples.

5. What are the strengths and weaknesses of humans when it comes to processing information?

6. What are the strengths and weaknesses of humans when it comes to making decisions?

7. In which areas of aircraft control do machines perform better than humans?

8. Argue in favor of or against the proposal that it would be safer to fly modern commercial aircraft without pilots.

9. Describe the disadvantages of automation.

10. What are the principles that should be adhered to when designing automated systems?

11. What is management by exception?

REFERENCES AND RECOMMENDED READINGS

Gibbons, A., Trollip, S.R., & Karim, M. 1990. The expert flight plan critic: A merger of technologies. *Educational Technology*, March 1990.

Green, G.G., Muir, H., James, M., Gradwell, D., & Green, R.L. 1991. *Human Factors for Pilots*. Aldershot, U.K.: Gower Publishing.

Hawkins, F.H. 1987. *Human Factors in Flight*, Aldershot, UK: Gower Technical Press.

Jensen, R.S. 1981. Prediction and quickening in perspective flight displays for curved landing approaches. *Human Factors*, Vol 23, No. 3

Lintern, G., Roscoe, S.N., & Sivier, J. 1990. Display principles, control dynamics, and environmental factors in pilot performance and transfer of training. *Human Factors, 32, 299-317.*

O'Hare, D. & Roscoe, S. 1990. *Flightdeck Performance: The Human Factor*. Ames: Iowa State University Press.

Wiener, E. L. & D. C. Nagel (Eds.). 1988. *Human Factors in Aviation*. San Diego, CA: Academic Press, Inc.

Glossary

A

ACCIDENT — An occurrence associated with the operation of an aircraft which takes place between the time any person boards the aircraft with the intention of flight and all such persons have disembarked, and in which any person suffers death or serious injury, or in which the aircraft receives substantial damage.

ACCOMMODATION — The process of changing the shape of the eye's lens to bring an image into focus.

AGL — Above ground level.

AIR DENSITY — The density of the air in terms of mass per unit volume. Dense air has more molecules per unit volume than less dense air. The density of air decreases with altitude above the surface of the earth and with increasing temperature.

AIR ROUTE TRAFFIC CONTROL CENTER (ARTCC) — A facility established to provide air traffic control service to aircraft operating on IFR flight plans within controlled airspace, principally during the enroute phase of flight. When equipment capabilities and controller workload permit, certain advisory/assistance services may be provided to VFR aircraft.

AIR TRAFFIC CONTROL (ATC) — A service provided by the FAA to promote the safe, orderly, and expeditious flow of air traffic.

AIRPLANE SIMULATOR — A replica of a specific airplane's instruments, equipment, panels, and controls. It must have both motion and visual systems.

AIRPLANE TRAINING DEVICE — A full-scale replica of an airplane's instruments, equipment, panels, and controls. It does not have to duplicate the appearance and performance of a specific aircraft and does not have to have motion and visual systems.

AIRPORT RADAR SERVICE AREA (ARSA) — Regulatory airspace surrounding designated airports where ATC provides radar vectoring and sequencing on a full-time basis for all IFR and VFR aircraft. Participation is mandatory, and all aircraft must establish and maintain radio contact with ATC.

AIRPORT SURVEILLANCE RADAR (ASR) — Approach and departure control radar used to detect and display an aircraft's position in the terminal area and to provide aircraft with range and azimuth information.

AIRPORT TRAFFIC AREA — The airspace within a horizontal radius of five statute miles from the geographical center of the airport and extending from the surface up to, but not including, 3,000 feet above the airport elevation. The airport must have an operating control tower for this area to be in effect.

ALTIMETER — The instrument that indicates flight altitude by sensing pressure changes and displaying altitude in feet.

ALTIMETER SETTING — The barometric pressure setting used to adjust a pressure altimeter for variations in existing atmospheric pressure and temperature.

ALTITUDE — Height expressed in units of distance above a reference plane, usually above mean sea level or above ground level.

ANGLE OF ATTACK — The angle between the airfoil's chord line and the relative wind.

ANGLE OF INCIDENCE — The angle between the chord line of the wing and the longitudinal axis of the airplane.

APPROACH CONTROL — A terminal air traffic control facility providing approach control service.

ATTENTION — The ability of a human to concentrate on one or more tasks.

AUTOMATIC DIRECTION FINDER (ADF) — An aircraft radio navigation system which senses and indicates the direction to an L/MF nondirectional radio beacon (NDB) or commercial broadcast station.

AUTOMATIC TERMINAL INFORMATION SERVICE (ATIS) — The continuous broadcast of recorded noncontrol information in selected terminal areas. Its purpose is to improve controller effectiveness and to relieve frequency congestion by automating the repetitive transmission of essential but routine information.

AUTOMATION — The process whereby machines (usually computers) take over tasks normally performed by humans.

B

BEARING — The horizontal direction to or from any point, usually measured clockwise from true north (true bearing), magnetic north (magnetic bearing), or some other reference point, through 360°.

BEST ANGLE-OF-CLIMB AIRSPEED — The best angle-of-climb airspeed (V_X) will produce the greatest gain in altitude for horizontal distance traveled.

BEST RATE-OF-CLIMB AIRSPEED — The best rate-of-climb airspeed (V_Y) produces the maximum gain in altitude per unit of time.

BLACK-HOLE EFFECT — An effect caused by a dark area in front of a lighted area, resulting in the illusion of being too high.

BLIND SPOT — An area on the retina in which no vision is possible due to the attachment of the optic nerve.

C

CALIBRATED AIRSPEED (CAS) — Indicated airspeed of an aircraft, corrected for installation and instrument errors.

CEILING — The height above the earth's surface of the lowest layer of clouds or obscuring phenomena that is reported as broken, overcast, or obscuration and not classified as thin or partial.

CENTER OF GRAVITY (CG) — The theoretical point where the entire weight of the airplane is considered to be concentrated.

CLEAR AIR TURBULENCE — Turbulence that occurs in clear air, and is commonly applied to high-level turbulence associated with wind shear. It is often encountered near the jet stream, and it is not the same as turbulence associated with cumuliform clouds or thunderstorms.

CLOSED-LOOP SYSTEM — A system which the operator has the ability to control once in motion, based on the continuous feedback of information.

COCKPIT RESOURCE MANAGEMENT (CRM) — The process of managing all resources for the benefit of flying an airplane safely. These resources may be on or off the airplane and include both humans and mechanical systems.

COLD FRONT — The boundary between two airmasses where cold air is replacing warm air.

COLOR BLINDNESS (DEFICIENCY) — A reduced ability of the eye to make color discriminations, particularly in low-light conditions.

COMPASS HEADING — A compass reading that will make good the desired course. It is the desired course (true course) corrected for variation, deviation, and wind.

COMPUTER-BASED TRAINING (CBT) — The use of a computer (usually a micro-computer) for teaching knowledge and skills. Also known as COMPUTER-BASED EDUCATION (CBE), COMPUTER-BASED INSTRUCTION (CBI), or COMPUTER-ASSISTED INSTRUCTION (CAI).

CONCEPT MAP — A graphic depiction of the relationships between ideas, information, or concepts.

CONTROL/DISPLAY COMPATIBILITY — The principle that controls and displays should move in the same direction as the motion of the system they control or indicate.

CONTROL ZONE — Controlled airspace from surface to base of continental control area. A control zone may include one or more airports, and it is normally a circular area with a radius of five statute miles and any extensions necessary to include instrument approach and departure paths.

CONTROLLED AIRSPACE — Airspace designated as control zone, airport radar service area, terminal control area, transition area, control area, continental control area, or positive control area within which some or all aircraft may be subject to air traffic control.

CONVECTION — The circular motion of air that results when warm air rises and is replaced by cooler air. These motions are predominantly vertical, resulting in vertical transport and mixing of atmospheric properties; distinguished from advection.

COURSE — The intended or desired direction of flight in the horizontal plane measured in degrees from true or magnetic north.

CROSSWIND — A wind which is not parallel to a runway or the path of an aircraft.

CROSSWIND COMPONENT — A wind component which is at a right angle to the runway or the flight path of an aircraft.

D

DENSITY ALTITUDE — Pressure altitude corrected for nonstandard temperature variations. Performance charts for many older airplanes are based on this value.

DEVIATION — A compass error caused by magnetic disturbances from electrical and metal components in the airplane. The correction for this error is displayed on a compass correction card placed near the magnetic compass in the airplane.

DISORIENTATION — The sense of not being oriented.

DISPLACED THRESHOLD — When the landing area begins at a point on the runway other than the designated beginning of the runway.

DISTANCE MEASURING EQUIPMENT (DME) — Equipment (airborne and ground) to measure, in nautical miles, the slant range distance of an aircraft from the DME navigation aid.

DOWNBURST — A strong downdraft which induces an outburst of damaging winds on or near the ground. Damaging winds, either straight or curved, are highly divergent. The sizes of downbursts vary from 1/2 mile or less to more than 10 miles. An intense downburst often causes widespread damage. Damaging winds, lasting 5 to 30 minutes, could reach speeds as high as 120 knots.

E

EMPTY FIELD (SKY) MYOPIA — A resting state or focus of the eye caused by lack of well defined objects in the visual field.

EXPECTATION — The mental set that causes a person to expect a particular event or stimulus to occur.

F

FINAL APPROACH — A flight path of a landing aircraft in the direction of landing along the extended runway centerline from the base leg or straight in to the runway.

FIXATION — The condition in which the mind is focused on one stimulus to the exclusion of all others.

FLIGHT SERVICE STATION (FSS) — Air traffic service facilities that provide a variety of services to pilots, including weather briefings, opening and closing flight plans, and search and rescue operations.

FLY-BY-WIRE — A control system in airplanes in which the control surfaces of the airplane are connected to the pilot's controls by electric wires, not cables or any other direct system.

FREEZING LEVEL — A level in the atmosphere at which the temperature is 32°F (0°C).

FRONT — The boundary between two different airmasses.

G

GLASS COCKPIT — A cockpit in which most of the instruments are electronically displayed on a computer monitor.

GROUND EFFECT — An usually beneficial influence on aircraft performance which occurs while you are flying close to the ground. It results from a reduction in upwash, down-wash, and wingtip vortices which provide a corresponding decrease in induced drag.

GROUND PROXIMITY WARNING SYSTEM (GPWS) — An alarm indicating the airplane is too close to the ground.

GROUNDSPEED — Speed of the aircraft in relation to the ground.

H

HANGAR TALK — Anecdotes about flying; often about getting out of perilous situations.

HEADING — The direction in which the longitudinal axis of the airplane points with respect to true or magnetic north. Heading is equal to course plus or minus any wind correction angle.

HORIZONTAL SITUATION INDICATOR — A navigation device that embodies both directional and displacement information.

HUMAN FACTORS — The study of how people interact with their environments.

HYPERVENTILATION — The excessive ventilation of the lungs caused by very rapid and deep breathing which results in an excessive loss of carbon dioxide from the body.

HYPOGLYCEMIA — Low blood sugar.

HYPOXIA — The effects on the human body of an insufficient supply of oxygen. (1) HYPOXIC: oxygen deprivation caused by a lack of oxygen in the lungs; (2) ANEMIC: oxygen deprivation caused by the blood's inability to carry oxygen; (3) STAGNANT: oxygen deprivation caused by an obstruction to blood flow; and (4) HISTOTOXIC: oxygen deprivation caused by an inability of the body to extract oxygen from the blood.

I

ILLUSION — (1) AURAL: The incorrect interpretation of spoken or aural information; and (2) VISUAL: the incorrect interpretation of visual information.

INCIDENT — An occurrence other than an accident, associated with the operation of an aircraft, which affects or could affect the safety of operations.

INDICATED AIRSPEED — The speed of an aircraft as shown on the airspeed indicator.

INDICATED ALTITUDE — The altitude shown by an altimeter set to the current altimeter setting.

INDUCED DRAG — That part of total drag which is created by the production of lift.

INSTRUMENT FLIGHT RULES (IFR) — Rules that govern the procedure for conducting flight in instrument weather conditions. When weather conditions are below the minimums prescribed for VFR, only instrument-rated pilots may fly in accordance with IFR.

INSIDE-OUT DISPLAY — A display that assumes that the platform on which it is based is stationary and that the world moves about it.

INSTRUMENT LANDING SYSTEM (ILS) — Electronic guidance system allowing appropriately equipped airplanes to land under IFR conditions.

INSTRUMENT METEOROLOGICAL CONDITIONS (IMC) — Weather conditions requiring flight under IFR.

INTERACTIVE VIDEO — A subset of CBT using both video and computer images for training.

INTERFERENCE — The condition when one source of information interferes with a person's ability to process another.

INTERNATIONAL CIVIL AVIATION ORGANIZATION (ICAO) — Body coordinating international regulation of civil aviation.

J

JET STREAM — A narrow band of winds with speeds of 50 knots and greater embedded in the westerlies in the high troposphere.

JUDGMENT — (1) COGNITIVE: the process of making intellectual decisions; (2) PERCEPTUAL: the ability to make visual decisions, such as height above the ground when landing. (Also see Pilot Judgment).

L

LAPSE RATE — The rate of decrease of an atmospheric variable with altitude; commonly refers to a decrease of temperature or pressure with altitude.

LATITUDE — Measurement north or south of the equator in degrees, minutes, and seconds.

LEANS — The condition when a rapid attitude correction in an airplane induces the illusion that the attitude has not been corrected, leaving the feeling of "leaning" in the direction of the correction, often causing an inappropriate secondary correction.

LOAD FACTOR — The ratio of the load supported by the airplane's wings to the actual weight of the aircraft and its contents.

LONGITUDE — Measurement east or west of the Prime Meridian in degrees, minutes, and seconds.

M

MAGNETIC COURSE — True course corrected for magnetic variation.

MAN-MACHINE SYSTEM — A collection of components, some of which are machines and at least one of which is a human, that work together to accomplish some purpose.

MANEUVERING SPEED (V_A) — The maximum speed at which full and abrupt control movements will not overstress the airplane.

MEAN SEA LEVEL (MSL) — The average height of the surface of the sea for all stages of tide; used as a reference for elevations throughout the U.S.

MICROBURST — A small downburst with outbursts of damaging winds extending 2.5 miles or less. In spite of its small horizontal scale, an intense microburst could induce wind speeds as high as 150 knots.

MILLIBAR — A unit of atmospheric pressure equal to a force of 1,000 dynes per square centimeter.

MNEMONIC — A technique for remembering and recalling information.

N

NATIONAL TRANSPORTATION SAFETY BOARD (NTSB) — The body charged with overseeing the safety of the United States' transportation systems. It undertakes accident investigations of airplane accidents.

O

OBSTRUCTION LIGHT — A light, or one of a group of lights, usually red or white, mounted on a surface structure or natural terrain to warn pilots of the presence of a flight hazard.

OPEN-LOOP SYSTEM — A system in which the operator has no control once in motion.

OUTSIDE-IN DISPLAY — A display that assumes that the world is stationary and that the platform on which the display is based moves about it.

P

PILOT ERROR — An action or decision of the pilot that if not caught and corrected could contribute to the occurrence of an accident or incident.

PILOT IN COMMAND — The pilot responsible for the operation and safety of an aircraft.

PILOT JUDGMENT — The mental process that we use in making decisions.

PILOT WEATHER REPORT — A PIREP is a report of meteorological phenomena encountered by aircraft in flight.

POOR JUDGMENT CHAIN — A series of judgment errors

PSYCHOMOTOR SKILL — A skill that requires both cognitive/perceptual and motor skills.

R

RADIAL — A navigational signal generated by a VOR or VORTAC, measured as a magnetic bearing from the station.

RECIPROCAL — A reverse bearing, opposite in direction by 180°.

RESPIRATION — The transfer of oxygen in the lungs to the bloodstream.

RESTRICTED AREA — Special use airspace of defined dimensions within which the flight of aircraft, while not wholly prohibited, is subject to restrictions.

RUNWAY THRESHOLD — The beginning of the landing area of the runway.

RUNWAY VISIBILITY VALUE — Visibility determined by a transmissometer for a particular runway.

RUNWAY VISUAL RANGE — An instrumentally derived value representing the horizontal distance a pilot will see down the runway from the approach end, based on either the sighting of high intensity runway lights or on the visual contrast of other targets, whichever yields the greatest visual range.

S

STANDARD ALTIMETER SETTING — An altimeter set to the standard pressure of 29.92 in. Hg, or 1013.2 Mb.

STANDARD ATMOSPHERE — A hypothetical atmosphere based on averages in which the surface temperature is 59°F (15°C), the surface pressure is 29.92 in. Hg (1013.2 Mb) at sea level, and the temperature lapse rate is approximately 2°C per 1,000 feet.

STRESS — The body's response to demands placed on it. (1) PHYSICAL: the physical response to stressors such as heat, noise, and vibration; (2) PHYSIOLOGICAL: the response to stressors such as fatigue, lack of sleep, and missed meals; and (3) EMOTIONAL: the body's response to social or emotional stressors, such as peer pressure or marital problems, fear, euphoria, anger, etc.

STRESSOR — Something that causes stress.

T

TERMINAL CONTROL AREA (TCA) — Controlled airspace extending upward from the surface or higher to specified altitudes, within which all aircraft are subject to the operating rules and pilot/equipment requirements as specified in FAR Part 91.

TRACK — The actual flight path of an aircraft over the ground.

TRAFFIC ADVISORIES — Advisories issued to alert a pilot to other known or observed air traffic which may be in such proximity to their position or intended route of flight as to warrant their attention.

TRAFFIC PATTERN — The traffic flow that is prescribed for aircraft landing and taking off from an airport. The usual components are the upwind, crosswind, downwind, and base legs; and the final approach.

TRANSITION AREA — Controlled airspace designed to contain IFR operations for aircraft traveling between the terminal and enroute environment. Beginning at 700 or 1,200 feet AGL, transition areas terminate at the base of the overlying controlled airspace.

TRANSPONDER — An electronic device aboard the airplane that enhances an aircraft's identity on an ATC radar screen.

TRANSPORTATION — The carriage of oxygen by the blood through the body.

TRUE AIRSPEED (TAS) — The speed at which an aircraft is moving relative to the surrounding air.

U

UNICOM — A nongovernment communications facility which may provide airport information at certain airports.

V

VARIATION — The angular difference between true north and magnetic north; indicated on charts by isogonic lines.

VESTIBULAR SYSTEM — The system of the inner ear that provides information concerning the body's orientation.

VICTOR AIRWAY — An airway system based on the use of VOR facilities. The north-south airways have odd numbers (Victor 11), and the east-west airways have even numbers (Victor 14).

VIGILANCE — The ability to detect a change in state or status.

VISIBILITY — The distance one can see and identify prominent unlighted objects by day and prominent lighted objects by night.

VISUAL FLIGHT RULES (VFR) — Rules that govern the procedures for conducting flight in visual conditions. The term "VFR" is also used to indicate weather conditions that comply with specified VFR requirements.

VISUAL METEOROLOGICAL CONDITIONS (VMC) — Weather conditions allowing flight under VFR.

VOR (VERY HIGH FREQUENCY OMNI-RANGE) — The primary electronic navigation system.

W

WIND CORRECTION ANGLE (WCA) — The angular difference between the heading of the airplane and the course.

WIND SHEAR — A sudden, drastic shift in wind speed, direction, or both that may occur in the vertical or horizontal plane.

WORKLOAD — (1) PHYSICAL: The amount of simultaneous physical activity performed by a human; (2) MENTAL: The amount of simultaneous mental activity performed by a human.

Index

A